SERVICES AND ECONOMIC DEVELOPMENT IN THE ASIA-PACIFIC

Services and Economic Development in the Asia-Pacific

PETER W. DANIELS
University of Birmingham, UK
&
JAMES W. HARRINGTON
University of Washington, USA

Routledge
Taylor & Francis Group

LONDON AND NEW YORK

First published 2007 by Ashgate Publishing

2 Park Square, Milton Park, Abingdon, Oxon OX14 4RN
711 Third Avenue, New York, NY 10017, USA

Routledge is an imprint of the Taylor & Francis Group, an informa business

First issued in paperback 2016

British Library Cataloguing in Publication Data

Services and economic development in the Asia-Pacific. -
(The dynamics of economic space)
1. Service industries - Asia 2. Service industries -
Pacific Area 3. Asia - Economic conditions 4. Pacific Area
- Economic conditions
I. Daniels, P. W. II. Harrington, J. W. (James W.). 1957 -
338.4´7´00095

Library of Congress Cataloging-in-Publication Data

Services and economic development in the Asia-Pacific / edited by
P. W. Daniels and J. W. Harrington.
    p. cm. -- (The dynamics of economic space)
Includes index.
ISBN 978-0-7546-4859-8 (alk. paper)
1. Service industries--Asia. 2. Service industries--Pacific area 3. Asia--Economic conditions
4. Pacific area--Economic conditions. I. Daniels, P. W. II. Harrington, J. W. (James W.), 1957

HD9987.A2S475 2007
338.4´095--dc22

                              2006037655

ISBN 13: 978-0-7546-4859-8 (hbk)
ISBN 13: 978-1-138-26267-6 (pbk)

List of Contents

List of Figures

List of Tables

List of Contributors

Harald Bekkers, Ph.D. (Amsterdam), currently works for the Swisscontact-KATALYST market development project, which aim is to develop markets for business services, increase enterprise competitiveness and (re)organize national and global value chains. Recent publications and contributions include: Bekkers, H. (2005), 'Between fixing and forecasting: provincial Ahmedabad brokered into a bridgehead for globalization from below', unpublished PhD thesis, University of Amsterdam. Bekkers, H. (2003), 'Growing dependence of public banking on private consultants for market expertise and risk management in India', *Asian Business and Management*, Vol. 2 (2), pp. 239–66. Bekkers, H. (2007), 'Brokering globalization from below in provincial Ahmedabad: The need for providers regulating intangible services and competing interests', in Visscher, S. and Muijzenberg, O. van den (eds), *Globalizing Cities, Local Practices: Analyses of Globalization from Below in Modern Asia, 1960–2000* (in preparation).

Pin-Hsien Chen is currently a first-year Ph.D. student in the Center for the Environment, University of Oxford. She examined the process of financial liberalization in her master thesis, and in her Ph.D. programme she will expand the scope of this research to incorporate financialization issues across the East Asian states.

Peter W. Daniels is Professor of Geography and Co-Director, Services and Enterprise Research Unit, University of Birmingham (UK). He has undertaken research on the service economy, especially producer services as key agents in metropolitan and regional restructuring at the national and international scale. Recent publications include: *Service Worlds: People, Organizations, Technologies* (with J.R. Bryson and B. Warf, 2004); *Service Industries and Asia-Pacific Cities: New Development Trajectories* (ed., with K.C. Ho and T.A. Hutton, 2005); *The Service Industries Handbook* (ed., with J.R. Bryson, 2007); *Knowledge-Based Services* (ed., with J.W. Harrington, 2006).

Peter Elliott is Senior Research Officer with the Department of Sustainability and Environment, State Government of Victoria, Australia. He has a BA (Hons) in Urban Planning from Victoria University, Melbourne and a MSc in Planning and Design in Economic Geography from The University of Melbourne. He is currently researching a number of themes centred on the economic geography of Melbourne and Victoria including the location and distribution of office employment, the supply and demand for industrial land across Melbourne, as well as continuing his work on graphic design firms.

Koshi Hachikubo is Associate Professor of Geography at Kanagawa University (Japan). He has undertaken research on industrial geography, especially the regional system of local industries. Recent research is published in *Knowledge, Industry and Environment* (Hayter and Le Heron, 2002).

James W. Harrington, Professor of Geography, University of Washington, Seattle WA. Professor Harrington has written on industrial location, service sectors, and regional economic development, primarily in North America. His current research focuses on occupational attainment, and on region-specific institutions for leadership and labor force development.

Noboru Hayashi is Professor of Geography, Nagoya University Graduate School of Environment and School of Informatics and Sciences (Japan). He has undertaken research on urban economic geography, especially retail and service activities as key agents in metropolitan and regional organization at the national and international scale. Recent publications include: *Urban Economic Geography*, 2002; *Modern Urban Geography*, 2003; *Regional Structure of Contemporary Canadian Cities*, 2004; *Geography of Urban Services*, 2005.

Jinn-Yuh Hsu is currently a professor in Geography at the National Taiwan University. He received his Ph.D. degree from the University of California at Berkeley. He is an economic geographer who specializes in research on high-technology and service-intensive industries and regional development in late-industrializing countries, particularly Taiwan. He is currently engaged in a research project to explore the evolution of financial liberalization in the East Asian developmental states, such as Taiwan and South Korea after the late-1980s transformation and since the neo-liberalization of the East Asian states became prominent in the early 2000s.

George C.S. Lin is Associate Professor and Head of the Department of Geography, University of Hong Kong. He is the author of *Red Capitalism in South China: Growth and Development of the Pearl River Delta* (University of British Columbia Press, Vancouver, Canada, 1997) and many articles. His research interests include the growth of services in Chinese cities, land management and land development processes, regional development in the Pearl River Delta, transnationalism, cross-border population mobility, and the geography of Chinese diaspora.

Shuguang Liu is Professor of Geography and Director, Regional Economic Research Institute, Ocean University of China (China). He has undertaken research on global value chains and regional industrial upgrading, especially maritime industrial upgrading at the national and international scale. Recent publications include: *Regional innovation system: theoretical approach and empirical study* *of provincial regions in China* (with C. Chen, 2003); *The role of new ICTs in the internationalization of firms: case study of Haier* (with W. Liu, 2003); *Progress in global value chains and regional industrial upgrading* (with H. Yang, 2005); *From Corporate standard to global standard: technological innovation and standard setting* (with G. Guo, 2006); *Global Value Chain Governance and Regional Maritime Industry Upgrading* (ed.).

Hideo Mori is Senior Researcher (Geography), Nippon Institute of Technology (Japan). He has undertaken the research on industrial geography, especially regional industrial complex in inner Tokyo. Recent research is published in *Knowledge, Industry and Environment* (Hayter and Le Heron, 2002).

Kevin O'Connor is Professor and Head of Urban Planning, University of Melbourne. His teaching and research explores the links between the service sector and the growth and internal structure of cities often using Australia and Melbourne as a case study. Recent books include: *The New Economic Geography of Australia: A Society Dividing* (2001) and *Melbourne 2030: Planning Rhetoric versus Urban Reality*. Papers published in 2005 include: 'Understanding mega-city development: a case study of Melbourne', 'Regional population and employment change in Australia 1991–2001: inertia in the face of rapid change?' and 'The importance of proximity in economic competitiveness: rethinking the role of clusters in local economic development'.

Yufang Shen is Professor of Geography and Deputy Director, Spatial Planning and Regional Economics Institute, and Associate Director, Yangtze Basin Development Institute, East China Normal University (China). He has undertaken research on the regional economy, especially urban and regional development and planning in metropolitan and regional restructuring process at the national and international scale. Recent publications include: *A Study on the Coordinated Economic Development between Shanghai and Other Areas alongside the Yangtze* (with W. Yang, 2001); *Investment, Development and Cooperation in the Yangtze Economic Zone* (2003); *Report on the Industrial Development of the Yangtze Valley* (with H. Zhang, 2003); *Policy Environment and Management Mechanism for Development of the R&D Centres Invested by Foreign Companies* (with H. Zhang and Z. Zhang, 2004); *Theories and Practice in Coordinated Development of the Regional Economy in China* (in press, 2006).

Atsuhiko Takeuchi is Emeritus Professor, Nippon Institute of Technology (Japan). He has undertaken research on industrial geography and industrial policy, especially regional industrial systems of Japanese machinery industries and industrial dynamics of metropolitan regions. Recent publications include: *Globalization and Dynamics of Industrial Region* (2005, Japanese); *An Economic Geography of Japan* (2005, Japanese); *Environmental Change and Industrial Complex* (2006, Japanese) and

in *New Economic Spaces: New Economic Geographies* (Le Heron and Harrington, 2005); *Knowledge, Industry and Environment* (Hayter and Le Heron, 2002).

Steffen Wetzstein is a doctoral candidate in the School of Geography and Environmental Science, University of Auckland. His thesis is on the 'Economic Governance for a Globalising Auckland?: Political Projects, Institutions and Policy', and is focused on Auckland's economic development after the neoliberal reforms of the 1980s. His broader research interests are in the areas of urban and regional governance, the role of the state and policy approaches to sustainable and competitive economic transformation in territorial contexts. He has undertaken a range of research and policy development projects for Auckland's local government. He published 'A research project as narrative: the making of a 'Creative Auckland' story' with Professor Richard Le Heron in the proceedings of the New Zealand Geographical Society twenty-second Conference, 2003.

Fiona F. Yang is Ph.D. Candidate of Geography, University of Hong Kong. Her research examines the growth and location of service industries in China from a political economy perspective. Publications include: 'Services and metropolitan development in China: the case of Guangzhou', in *Progress in Planning* (2004).

Shuang Yann Wong is Associate Professor of Geography, Humanities and Social Studies Education, National Institute of Education, Nanyang Technological University, Singapore. She has undertaken research on the economic development and industrial restructuring in Southeast Asian countries at the national, regional and international scale. Recent publications include: 'Cross-national ethnic networks in financial services: a case study of local banks in Singapore', in *Linking Industries Across the World* edited by Claes G. Alvstam and Eike W. Schamp, 2005; 'State governance: regulatory processes and entrepreneurship: Singapore's concentrating banking sector', in *New Economic Spaces: New Economic Geographies*, edited by Richard Le Heron and James W. Harrington, 2005.

Preface

. . . .

As the role of services in economies, from the global to the local, increases and diversifies, significant changes to the way in which they are produced and consumed are under way. Continual change and evolution is one of the primary features of all economies; it is driven by the forces of competition, technological innovation, and new forms of management, organization and work. Symptomatic of these changes is the transformation of economies from manufacturing to services. This does not necessarily mean that all economies are shifting away from manufacturing into services in the same way or at the same rate, but there is an ongoing 'blurring' of the long established distinction between the two. Extensions to the global reach of services and their engagement with the processes of globalization has been accompanied by an increasing spatial division of labour. This on-going transformation has stimulated new support functions that feed into the production process (manufacturing and services) as well as increasingly driving a new economic geography in which service industries are one of *the* key players. Services industries are also important for local economic growth in different national or urban settings, with substantial and differential impacts in developed, developing and transition economies.

This provided the context for the annual residential conference of the International Geographical Union's Commission on the Dynamics of Economic Spaces held at the University of Birmingham in August 2004. Under the rubric of 'Service Worlds: Employment, Organizations, Technologies', contributors explored, amongst others, ways of understanding changes in the production and consumption of services and the impacts on economic development, the globalization of services and the dynamics of city/regional systems, internationalization of advanced business and professional services, the embeddedness of service enterprises in local economies and their ability to generate local social capital, and the role of information technology the regional/global redistribution of service work. Peter Daniels and Michael Taylor served as the organizers and conveners of the meeting.

This is the second volume derived from the conference proceedings. Each reflects a common strand that links several of the contributions; for the first volume this (Harrington, J.W. and Daniels, P.W. (2006), *Knowledge-based Services, Internationalization and Regional Development*) was the role of knowledge and knowledge-based services while for this volume the unifying dimension is the economies of the Asia-Pacific as a theatre for exploring the relationship between service industries and economic/urban restructuring, including the effects of regulation and public policy.

We are indebted to all the contributors for patiently awaiting our initial decisions about which papers to include in this volume and their subsequent prompt responses to our requests for changes to their manuscripts or the inclusion of additional material. The prompt publication of volumes of this nature relies heavily on the cooperation of the contributors: we gratefully acknowledge their generous support for this project.

P.W. Daniels   J.W. Harrington
University of Birmingham, U.K.   University of Washington, Seattle, U.S.A.

Chapter 1

Services and Development in the Asia-Pacific: Introduction and Overview

Peter W. Daniels and James W. Harrington

Introduction

The early twenty-first century has witnessed growing attention to the contribution of service industries to the processes of urban, economic and social development in the Asia-Pacific (Daniels et al., 2005; Hutton, 2001, 2004). To this literature, this volume brings a particular focus on the role of public policies (governmental and civil) at a variety of scales and their geographical outcomes at both national and urban levels. The value of service industries to the Asia-Pacific economies is often underrated; manufacturing still attracts most of the political and administrative energy when it comes, for example, to designing tax, trade and support policies (APEC, 2002). However there are clear signs that things are changing. This relative inattention has not caused the service sector to stagnate, though it has probably undermined the sector's ability to fulfil its potential to deliver the economic benefits that it might.

Although industrialization was the dominant development paradigm amongst many Asia-Pacific nations for much of the second half of the twentieth century, during the 1990s and at the beginning of the twenty-first century services have become more deeply embedded in the region's economic growth and structural change. Rapid tertiarization is fundamentally restructuring national and regional economies, especially in the region's major cities (Daniels et al., 2005; Douglas, 2002; Olds, 2001; Hutton 2001). In Shanghai, for example, service-producing industries have grown rapidly and, although still some way behind comparable cities in Europe or North America, accounted for some 51 per cent of its total economy in 2005 (compared to 40 per cent for China as a whole) (*China Daily*, 2006). There is an increasing recognition of the importance of services, in part because they have become increasingly interconnected with the goods sectors as a way of enabling the latter to remain competitive (Harrington and Daniels, 2006). Activities such as finance, business, information, and creative services have become 'a defining feature of a new developmental trajectory with strategic linkages to larger processes of change' (Hutton, 2005, 54). It is now accepted that services play a critical role in determining both the quality and speed of the process of economic development, and that a competitive economy cannot exist without an efficient and technologically advanced service sector (APEC, 2003).

The Key Role of Knowledge-Based Industries

Instrumental to these trends has been the contribution of knowledge-based industries. A number of these can be classified as essential, including producer services which are now decisive for the operation of more advanced and 'flexible' production regimes. Countries depend on banking and financial institutions as the key intermediaries for flows of capital, and to facilitate their integration into a global market. Information technology services are essential both for the development of a knowledge-based economies in the region as well as enabling international telecommunications and connectivity which are complemented by physical communications services such international airports, ports, shipping and allied industries connecting the system of global gateway cities – of which there are several in the Asia-Pacific region. Finally, creative and applied design services are amongst the basic industrial underpinnings of the cultural economy of cities in the Asia-Pacific (and elsewhere).

Services are seen one of the instruments for metropolitan modernization and reconfiguration that incorporate globalization strategies that are put in place and promoted by central and regional governments.[1] Many Asia-Pacific economies have world-class IT infrastructure; two of the top three economies worldwide for mobile phones *per capita* are located in the region (Chinese Taipei and Hong Kong) while Japan and the Republic of Korea lead the world in commercial deployment of 3G networks. At the same time, trade in services has also expanded rapidly with the region accounting for nearly half of the world's volume of trade in commercial services in 2004 (WTO, 2005).

Exemplars of the Shift to Services

There are various ways in which these trends can be illustrated. Take the example of Japan's general trading companies (*sogo shosha*) which have traditionally accounted for a large share of its total trade but which have witnessed a 40 per cent decline in that share over the last 20 years. There are now just seven *sogo shosha* out of 11,000 trading companies in Japan, although they continue to make a disproportionate contribution to the trading activities of Japan's major corporate groups (*kieretsu*). However, their decline has been accompanied by a notable shift of their overseas investment activities into services (UNCTAD, 2004). Rather than confining themselves to trade-related activities they have diversified into producer services such as business promotion, research and information, market development, group management, risk management, logistics, and finance. *Sogo shosha* also invest in many industries but here, too, the targets are changing: prior to 1980, 46 per cent of foreign affiliates (1,338 firms) were in manufacturing with only 31 per cent in commerce (services). According to the 2003 annual reports of the five largest *sogo*

---

1   A recent example is the China Wuxi World Cities Service Industries Convention held in Wuxi in November 2005 and organized, amongst others, by the China Council for the Promotion of International Trade and the Coalition of Service Industries (US).

shosha, the balance had reversed with 49 per cent of the 660 foreign affiliates listed as their principal subsidiaries and associated companies in commerce and 28 per cent in other services (after UNCTAD, 2004, footnote 3). In addition, manufacturing no longer constitutes the largest single sector of their FDI portfolio: foreign affiliates in services represent 69 per cent (of which half is in commerce). Of the 139 foreign affiliates established by the five largest *sogo shosha* in 2000–2002, 82 per cent were in services (40 per cent in commerce and 42 per cent in other services) (UNCTAD, 2004).

While the levels of employment creation in host countries as a result of services FDI are not necessarily higher than those associated with manufacturing FDI, the potential for job creation is also growing with the rise of other export-oriented services such as tourism (UNCTAD, 2004). Travel and tourism services generate significant economic growth, especially for those economies in the region that possess the appropriate factor endowments, as well connecting cities to cities and countries to countries inside and outside the region. FDI in these services generates more employment per dollar than FDI in location-bound services.

The growth of FDI in research and development (R&D) activities is another indicator of the Asia-Pacific region's expanding engagement with services (UNCTAD, 2005). Largely confined to developed economies until quite recently, R&D expenditures in China, Singapore, Hong Kong (China), Malaysia and the Republic of Korea by majority-owned foreign affiliates of US transnational corporations (TNCs) increased from US$400 million in 1994 to more than US$2.1 billion in 2002. In China alone the number of inwardly invested R&D establishments increased by 700 between 1995 and 2005, from a standing start. More than half of the 300 largest R&D spending firms in the world now have R&D facilities in Singapore, India or China (UNCTAD, 2005). Countries in the region that are particularly successful at achieving significant levels of R&D activity are proactive in the formulation and implementation of government policies aimed at enhancing their 'created assets'. They do this by vigorously promoting imports of know-how, people, capital and technology alongside domestic strategic investments in human resources, specialized infrastructure such as technology parks, strengthening and regulating financial institutions, or introducing targeted incentives for attracting knowledge-intensive investment. An important lesson is that policies to strengthen education, competition, FDI, and business innovation, as well as policies targeting the needs of specific industries and smaller firms, are effective, especially if they are well coordinated and adopt a long time horizon. An example is the Republic of Korea where some 140 foreign affiliate research institutes, including one of Microsoft's four overseas research centres and the R&D establishments of Intel, Motorola, Philips and Siemens, had all been opened by the end of 2004 (UNCTAD, 2005). As the synergy between national policies for creating sustainable knowledge-based clusters and inward investment in R&D deepens it also enables firms based in the Asia-Pacific to invest in their own overseas R&D activities. According to UNCTAD (2005) in 2004 there were 60 R&D centres in other part of the world owned by firms based in the Republic of Korea; Chinese companies operate at least 80 R&D units in other parts of the world and it is well documented that software firms based in India are now very actively engaged in creating a direct R&D presence in the US and elsewhere.

In addition to the direct impact of services FDI on employment, it is crucial not to overlook the indirect effects, the greater availability and better quality of producer or intermediate services as a result of FDI stimulates production in client industries (including additional jobs). In host countries where supplier industries of international standard exist or can be developed, production and employment in upstream industries can also increase. It therefore makes good sense for the Asia-Pacific economies to implement policies and initiatives aimed at diversifying and improving the quality of knowledge-intensive services. However, it is important to weigh the positive versus the negative impacts of engaging with transnational services (TNS).

On the positive side, the TNS inject additional capital and encourage the restructuring and rehabilitation of competing domestic activities, thereby strengthening the local economy and making it more resilient to external competition. TNS may directly introduce new service technology and products which will also have the effect of raising the quality and diversity of domestic service firms, making them more competitive (including pricing effects). The entry of TNS can also prompt an intensification of the market infrastructure in the form of improved legal, auditing and other regulatory standards which create more stability within the host economy. On the negative side, the arrival of TNS can weaken the domestic service economy, especially if the there is a weak regulatory framework that allows excessive concentration of an activity in ways that discourage market competition. TNS may also 'cherry pick' host country (as well as other international) clients, making it more difficult for domestic service firms to be exposed to high level client expectations which, in turn, makes it more difficult to provide a service offer that can equal or exceed that of their competitors. If domestic services cannot compete with TNS they may also be tempted to pursue more risky commissions or business which, if delivered unsuccessfully, will dilute their reputational capital. TNS are able to relocate some of their factors of production (including personnel) quickly, perhaps as a result of an economic or a political crisis for example, making for uncertainty about the stability and development trajectory of the host economy which is, in any event, also exposed to the tendency for repatriation of profits/fee income which negatively impacts the balance of payments.

Local, Regional and International Interventions

The balance of the risks and opportunities associated with raising the quantity and quality of services in national economies rests partly on the interventions of local, regional and national governments into internal and international service flows (see for example Stern, 2001). The strategic management of services trade liberalization has become integral to international trade negotiations following the adoption of the first multilateral trade agreement for services (The General Agreement on Trade in Services, GATS) which entered into force in January 1995 as a result of the Uruguay Round of trade negotiations.[2] The GATS is intended to contribute to trade expansion 'under conditions of transparency and progressive liberalization and as a means of promoting the economic growth of all trading partners and the development of developing countries'. For many Asia-Pacific economies, an important part of their comparative advantage resides in parts of the service sector such as transportation and tourism. Such comparative advantage can be utilised in the interests of economic growth and development if access to the markets of their trading partners can be managed in a transparent and mutually equitable way e.g. in the context of trading rules and regulations that allow equal treatment of the partners.

There is a continuing debate about whether the GATS is really necessary, because many services are considered to be national domestic activities or primarily in the domain of national government ownership and therefore not open to the application of trade policy concepts and instruments (WTO, 2006). But other large sectors, such as telecommunications, have undergone fundamental technical and regulatory changes in recent decades, opening them to private commercial participation and reducing or completely removing existing barriers to entry. The Internet has created a completely new range of internationally traded services such as e-banking, customer support services, tele-health and diagnostics, or distance learning that were unknown as recently as two decades ago. It has also removed distance-related barriers to trade in professional services, for example, that had disadvantaged suppliers and users in remote locations. The expansion of electronic networks has not only opened possibilities long-distance services trade but has also allowed the unbundling of the production and consumption of information-intensive services activities such as research and development, computing, inventory management, quality control, accounting, secretarial, marketing, advertising, distribution, and legal services that has formed part of the outsourcing/offshoring debate (Abramovsky et al., 2004; Blunden, 2004).

This brief introduction highlights the tensions between the global and the local processes shaping the pace and trajectory of extant and emerging service economies in the Asia-Pacific region. There is a strong desire to maintain control over changes driven or stimulated by service-producing activities which are prone to take place quickly, and possibly to the detriment of the host economies and cities, while recognizing that over-regulation may limit some of the benefits that also arise. The prospect of losing valuable service-related investment to other places, especially cities, in the region is but one symptom of the competition to be regionally prominent but also globally visible.

---

2   This was almost half a century after the entry into force of the General Agreement on Tariffs and Trade (GATT) of 1947, the GATS' counterpart in merchandise trade.

Overview of the Contents of this Volume

The contributions to this volume reflect these themes and have been grouped into three sections. The next four chapters examine aspects of the effects of the service dimension on the dynamics of economic spaces, especially in and around major cities.

Graphic design services are not only an important part of the advanced service sector in the leading cities of the region, they have also increased in number as a result of increasing demand linked to advertising and multi-media expansion in the wake of improved ICT. Peter Elliott and Kevin O'Connor (Chapter 2) contend that research on advanced services such as graphic design has demonstrated their propensity to take advantage of the agglomeration economies available in large metropolitan areas, but has overlooked the extent to which the resulting clusters of these activities are dynamic over time. Chapter 2 extends that understanding through an analysis of clusters of graphic design firms are identified in metropolitan Melbourne. Changes in their location, size, and density are traced over a 20 year time horizon from 1981 to 2001. They demonstrate continuity, in that the initial graphic design cluster in 1981 was still present by 2001, alongside change represented by growth in the number of firms in the industry accompanied by re-locations that have resulted in new clusters around the CBD as well dispersal of firms to locations in middle and outer suburbs. The intra-urban dynalmics of the graphic design cluster/s are accounted for by measuring the influence of variables such as size of firm, age of establishemnt and types of graphic design activity.

One symbol of the rapid growth of the economy of China in recent years is the boom in car sales and this has induced innovations in the sale and marketing of cars that have significant impacts on land use and urban form. This is illustrated by Yufang Shen (Chapter 3) using the case example of Shanghai where the new kind of car-selling market and market place represent a break with past practices; Shen calls this 'the third form', a new producer service linked to the car-making industry. Beginning in 2003, two giant car-selling service cities along with three car-selling service avenues are being constructed within the central area of Shanghai. The Shanghai International Car City is the largest of these projects which involves investment of several billion yuan. It includes almost all the service-producing businesses that are now essential to successful car production: marketing, car-related logistics, R&D, commercial, financial, lease, accounting, insurance and legal services, entertainment (particularly the F1 Racing Festival), food and shopping. The background and nature of the 'third form' of the car-making producer service industry is explored by way of informing an understanding of the location factors and main characteristics of the Shanghai International Car City. The spillover effects of the car city concept in relation to car producers in the vicinity and its more general impact on local economic development in Shanghai are examined, including a comparison with similar approaches to car marketing and the development of consumer awareness that have been used in the United States and in some EU countries such as Germany.

It has been gradually accepted that technological innovation by service-producing firms, especially knowledge-intensive business services (KIBS) is at least as important as the equivalent activity by manufacturing firms. As economies in the region shift towards services this becomes significant for competitive advantage, not least for the metropolitan areas where KIBS tend to agglomerate. In Chapter 4 Ji-Sun Choi explores the attributes of knowledge intensive business service (KIBS) firms in metropolitan Seoul compared with similar firms located in other parts of South Korea. Because KIBS tend to be concentrated in a few large cities it is assumed that their innovative capacity is likely to be superior to that of KIBS located in provincial areas. The major findings of this study suggest some policy implications in terms of spatial as well as industrial policies.

There is change of scale in Chapter 5 where Noboru Hayashi examines the locational factors determining the location of software services across Japan. Both temporal trends and spatial patterns are explored using survey data collected by the Ministry of Land, Infrastructure and Transport. In addition to demonstrating the agglomeration of software services and its close relationship with the urban hierarchy, the intra-urban location of these services is also illustrated in considerable detail for the major cities, including Tokyo, Osaka, Yokohama and Fukuoka. Location quotient based on the number of establishment and population in each prefecture indicated that Tokyo, Osaka, and Fukuoka had more establishments than expected by their population. The majority of businesses providing software services are small; more than half of establishments, for example, only occupy office floorspace of up to 50m². Accessibility to clients, the transportation network and especially proximity to rail stations are shown to be the most important factors inflencing the choice of location. The services available within office buildings, as well as the quality of the working environment, are also important.

The chapters that make up the second part of the book stand back from the evolving distribution and location behaviour of some important advanced or producer services; they focus more on the effects of the local and national regulatory environments on the growth of service generally or on specific, targeted service activities.

There is a now a rich body of theoretical literature that helps to explain the rise of services. Factors such as rising incomes, technical progress, and late capitalism have shed light on the growth dynamics of services but Yang and Lin (Chapter 6) suggest that this does not necessarily assist the understanding of the distinctive attributes and processes associated with the economic tertiarization now taking place in China. They seek to interpret the fast-growing tertiary sector in China since 1978 not only as a reflection of increasing *per capita* income and technical progress, but also as a product of political and social considerations shaped by the actions of the state. They argue that this is crucial to an understanding of the expansion of services in a socialist economy. In their case study of Guangzhou they investigate the role of services in metropolitan development in an effort to shed light on China's urban development more generally. The phenomenal expansion of service activities in Guangzhou does reflect rising disposable incomes, but also incorporates government intervention in the interests of social objectives. The service sector has become a new driving force behind urban population mobility and land use transformation. Their evaluation of state engagement and strategies toward the growth of the service sector suggests that established explanations for the growth of services need to be reconsidered in the light of the experience of a transitional socialist economy and, in particular, the growth of services in China's metropolitan areas.

The developmental state model is often used to explain the success of Singapore and government-linked companies (GLC) play a significant role. Shuang Yann Wong (Chapter 7) makes a detailed analysis of Singapore Airlines (SIA) as a GLC, now partially privatized, that is a pillar of the nation's services sector. It not only supplies key air transport services, it also support others key activities for Singapore, especially international tourism which is another important source of national revenue. In the wake of intensified competition in both market liberalization and innovation in air transport technologies, there have been doubts about the ability of SIA to remain the top airline service exporter in the region (South East Asia). SIA has effectively overcome the constraints of space and a small domestic market by optimizing benefits from the country's strategic location and well established linkages with the domestic, regional and global economies. Attention is drawn to forces beyond the power of the state, such as regional financial crisis in 1997, the outbreak of SARs, international terrorism and bird flu that have stunted the airline's growth. SIA was plunged into heavy deficits and the problems of employment and economic adjustments began. It is argued that the developmental state model may not be adequate for explaining the future for Singapore using the future roles, challenges, problems and constraints faced by SIA as it strives to sustain its competitive advantage to support the argument.

The strategic restructuring of multinational enterprises through worldwide alliance and localization in guest countries reinforces the position of large corporations in the governance of global value chains. This occurs at the expense of small and medium sized enterprises (SMEs) that may have served as the local suppliers of large corporations in host countries. With industrial upgrading in host countries of East Asia, there is need for SMEs to engage in internationalization and this is indeed occurring in the case of firms from Japan and South Korea that have set up operations in neighboring countries of the Asia-Pacific and have even gone as far as to Eastern Europe. Recent studies in China show, however, that an incubation process for foreign SMEs moving to host countries is needed if they are to successfully re-join the global value chains in the new milieu. Shuguang Liu and Guogang Ren (Chapter 8) use the Sino-Korean joint project known as *Window Korea* as a case to study for assessing the potential and future functions of international business incubator in the process of Korean SMEs moving to China *via* the nearest location on the Shandong peninsula. Their initial results indicate that the special business incubation system used for this project is essential, at least for Korean auto-part and electronic part SMEs, for the process of local survival in the host economy and their ability to rejoin worldwide value chains.

The basic regional economic policy of Japan since the 1960s has been designed to promote the dispersion of industries from the metropolitan areas to provincial areas. However, despite huge investment, there has not been a drastic change of the regional economic system. Effective regional industrial policy should be grounded on the physical, historical, socio-cultural conditions of the designated area and be related to the nature of the industry itself. This shaped the thinking behind the 'One village one product' project that seeks to achieve local economic growth by harnessing local resources and their latent potential explored by Atsuhiko Takeuchi, Koshi Hachukibo and Hideo Mori (Chapter 9). They investigate the formation and development of the *shochu* industry in Ohita prefecture and secondly the service-informed process by which this industry has successfully outgrown the *sake* industry in Japan. Continual transformation of the production system, including improved marketing and related services such as advertising and promotion, is shown to be crucial to the re-invigoration of an indutry that was in danger of being eclipsed by foreign imports of whisky, notably from the UK (Scotland).

The final section of three chapters illustrates the challenges faced in countries that are anxious to modernize their services or, in the case of Auckland, to maximize the interface with the global economy from a marginalized geographical position.

In Chapter 10 Jinn-Yuh Hsu and Pin-Hsien Chen tackle the issue of uncertainty in the development of the financial service industry (FSI) in Taiwan. In spite of being praised as a model of developmental state, the Taiwanese government was very conservative in its financial policy, and usually prioritized stability over growth in its decision making. As a result, it took an ambivalent attitude toward the opening of a futures market, which was believed to be highly speculative and subversive to the financial system. However, an underground futures market existed *de facto* and the state was forced to legalize the new industry. Even so, instead of following the big push model, it adopted a gradualist approach and opened the trading market step by step. This created a market dominated by domestic firms and little foreign capital participation; it was characterized as decentralized and price sensitive to service charges, and at the same time, required the brokers to stitch the knowledge gap for the individual investors, rather than the institutional ones, for such complex financial transactions. Under such circumstance, it was the extent of social networks, rather than the sophistication of trading skill that was critical for the business. The reluctant state policies have led to a decentralized market structure which has shaped an underdeveloped professional community in Taiwan's futures trading industry, thus limiting its contribution to the expansion of the FSI.

Cities in the Asia-Pacific that are geographically remote from the force of global economic integration face particular challenges in their efforts to integrate and enhance their engagement. Auckland's economy is domestically and consumption oriented. And Wetzstein (Chapter 11) suggests that promoting intentional connections between local and global economic processes will be problematic. Facilitation will probably need to be customized but tapping into, building and maintaining particular networks will certainly be an important aspect of economic governance. The mobilization of business interests is an important part of the process but the central state remains the main policy actor, in relation to 'stretching' and deepening participation if global networks of governance and how these are translated into effects on regional and urban processes in Auckland. His analysis shows that governing capitalist development remains a challenge in which inequalities are never far away, citing the example of 'non-exporters' or firms in the tourism and property sectors that are overlooked in the incentives available from central government to act globally. The key question is whether 'globally connecting' local infrastructure and business can be influenced by the state? It seems, from the Auckland experience to date, that the answer is uncertain and its greater incorporation in the Asia-Pacific region and the wider global economy remains in doubt.

Finally, Harald Bekkers (Chapter 12) shows how service providers act as 'brokers' in the process of economic globalization. They are able to re-embed exogenous best practices in a local context; thereby contributing to context-based modernization and the development of indigenous competitiveness in the face of global market competition. Private consultants lead this brokerage role in Ahmedabad, western India because institutional and/or public service providers are few and far between. He explores how the market for these private consultants is organized, distinguishing between layers, different networks and strategies through which consultants seek to find clients, and identifies 'best practices' for brokering exogenous, global best practices to a local business community. One finding is that Ahmedabad has a consultancy gap; consultants offer an intangible product for which a good reputation is essential so to be seen to act professionally is paramount. Consequently, they operate in limited circles and target the same set of medium-sized companies, leaving the much more numerous very small companies in Ahmedabad without any useful access to consultancy services. The second finding is that those consultants who are not totally preoccupied with the 'reputation game' are best placed to perform the brokerage role because they are willing to 'compromise' on their professional appearance and can therefore bridge the gap between the professional and the entrepreneurial domain.

Conclusions

This volume focuses on the role of governmental *and* civil policies at a variety of scales and their geographical outcomes at national *and* urban levels. State and civil actors are motivated to develop service sectors by a drive for modernization, a desire to enter international markets, to compete successfully with TNCs operating in their host countries, and/or to develop export earnings. Our key premise is that raising the quantity and quality of services in national economies depends partly on the interventions managed by local, regional and national governments, by actors within the sectors (especially in the absence of government policies), and international trade arrangements and economic conditions.

Through the contributions which follow, we find that spatial patterns of sectoral growth depend on the size and age structure of the sector and urban zoning and infrastructure policies. In a situation of rapid economic growth, a specific set of services (e.g., services supporting automobile production and marketing in metropolitan Shanghai, KIBS in South Korea, or software development in Japan) may grow up and establish strong agglomerative tendencies to support each other and their clients. Can governmental policies have substantial effects on these tendencies? Specifically, can infrastructure, zoning, and regulatory policy force dispersion of these sectors without harming the effectiveness of the services they develop and provide?

In a political-economic system where the state carries such great influence as in China, the effects of public regulation of services are readily apparent. Some Chinese provincial and local governments have deliberately emphasized services as a key component of modernization, with resultant differences in provincial or metropolitan rates of service-sector growth. Actions by the domestic state can influence the outcomes of inward FDI in services, as foreign-based operations seek to gain entry into global value chains from their host-country locations. However, for a sector as explicitly international as an international airline (SIA, in Chapter 7), *international* economic and regulatory exigencies have at least as much influence on sectoral outcomes as do the actions of the internal state.

State inattention or ambivalence to the development of a service sector (e.g. futures markets in Taiwan's financial services industry) can indeed slow the sector's development. However, it can also create a sector whose practitioners are more dependent on local connections and networks, for good (cohesiveness) and ill (lack of international competitiveness). Bekkers uncovers an analogous process in Ahmedabad, where many management consultants focus on a set of larger clients in order to maintain the prestige that is an important part of their marketing strategies.

Services growth is not a 'naturalized' process, proceeding at pace across the landscape. Besides the obvious importance of the urban hierarchy, and the greater geographic concentration of services in countries with a primate urban center, the rates and forms of service sector growth depends on explicit national and subnational policies, or on actions by corporate and professional networks in the absence of explicit policies.

References

Abramovsky, L., Griffith, R. and Sako, M. (2004), *Offshoring of Business Services and its Impact on the Economy*, London: Advanced Institute of Management Research.

APEC (2002), *The New Economy in APEC: Innovations, Digital Divide and Policy*, Singapore: APEC Economic Committee.

APEC (2003), *Drivers of the New Economy: Innovation and Organizational Practices*, Singapore: APEC.

Blunden, B. (2004), *Offshoring: the Good, the Bad and the Ugly*, Apress: Berkeley, CA.

Chinadaily.com.cn (*China Daily*) (2006), 'Local GDP revisions show service sector disparities', Monday, January 2. Retrieved August 23, 2006, LexisNexis™ Academic.

Daniels, P.W., Ho, K.C. and Hutton, T.A. (2005), *Service Industries and Asia-Pacific Cities: New Development Trajectories* (RoutledgeCurzon Studies in the Growth Economies of Asia), London: RoutledgeCurzon.

Douglass, M. (2002), 'Globalization, intercity competition and the rise of civil society: towards livable cities in Pacific Asia', *Asian Journal of Social Sciences*, 3(1), 131–152.

Harrington, J.W. and Daniels, P.W. (2006), 'International and regional dynamics of knowledge-based services', in J.W. Harrington and P.W. Daniels, op cit., 1.

Hutton, T.A. (2001), *Service Industries and the Transformation of Asia-Pacific City Regions*, Centre for Advanced Studies Research Paper Series: National University of Singapore.

Hutton, T.A. (2004), 'Service industries, globalization, and urban restructuring within the Asia-Pacific: new development trajectories and planning responses', *Progress in Planning*, 61(1), 1–74.

Hutton, T.A. (2005), 'Services and urban development in the Asia-Pacific region: institutional responses and policy innovation', in Daniels, P.W. et al., op.cit, 54.

Olds, K. (2001), *Globalization and Urban Change: Capital, Culture and Pacific Rim Mega-Projects*, Oxford: Oxford University Press.

Stern, R.M. (ed.) (2001), *Services in the International Economy*, Ann Arbor: University of Michigan Press.

UNCTAD (2004), *World Investment Report 2004: The Shift Towards Services*, Geneva: UNCTAD.

UNCTAD (2005), *World Investment Report 2005*, Geneva: UNCTAD.

WTO (2005), *International Trade Statistics 2005*, Geneva: WTO.

WTO (2006), 'The General Agreement on Trade in Services: An Introduction', Document 3776.4, March. Geneva: WTO.

PART I
Dynamics of Economic Spaces:
The Services Dimension

Chapter 2

Change in the Agglomeration of Service Firms in a Metropolitan Area: A Case Study of Graphic Design Firms in Melbourne 1981–2001

Peter Elliott[1]* and Kevin O'Connor[2]

Introduction

There is a well established understanding that related economic activities locate close to one another in parts of cities. Beginning with Marshall's observations on industrial areas in the UK (1920), and refined and expressed in the seminal studies of Wise (1949) in Birmingham, the idea of industrial districts was developed to capture that understanding. Much of the research into this concept had a manufacturing focus but subsequent studies (Piore and Sable, 1984, Scott, 1988, Saxenian, 1990) have shown that it was not just the difficulties associated with the physical movements of goods and people that made related functions locate close to one another but also the effect of local networks of technical skills, social, cultural and personal contacts of both managers and labor.

Some new circumstances have arisen which may modify the co-location of firms. At the level of physical goods, the quality and capacity of the logistical services that integrate air, sea and land transport have changed the priorities for location in most industries. That perspective is captured in large part in Dicken's (1998) idea of a 'global shift' in production. At one extreme, the research on the computer industry in South East Asia (Bowen et al., 2002) shows that global systems of component supply, and even management skill, are expressed in production systems that span continents (although the activity in the sector is still clustered at the local scale in a few sites within those continents). In another context, dispersal from the central and inner city to suburban and ex-urban industrial estates has become common, although clusters of related activities can still be found in these areas.

1   Department of Sustainability and Environment, State Government of Victoria. *The views expressed in this paper are those of the author and should not be regarded as representing the views of the Victorian Government or the Department of Sustainability and Environment.

2   *Urban Planning*, The University of Melbourne.

For advanced services, it has been suggested that the low cost and high capacity of telecommunication systems provide something similar to the locational freedom created by logistics services. In effect what can be seen as 'weightless products' (Quah, 1996) can be moved easily and quickly using information and communication technologies which may, in turn, free-up activity from locational constraints, usually associated with sites in the centres of cities (Quah, 1996). In fact O'Brien (1992) heralded the 'end of geography' in an analysis of the possible future location of banking centres, and Cairncross (1997) has suggested that distance no longer exerts a tyranny on location.

While the output of some producer services do exhibit the characteristics of a weightless product, there are elements in their production process where the old notions of spatial proximity are still relevant (Sassen, 2000). In short, it is the process of creating products, not necessarily the transport of the products themselves, that promotes the agglomeration of producer services firms in cities. These agglomeration factors include traditional elements such as access to a infrastructure and a skilled workforce (Scott, 1999) as well as less tangible elements such as the amenity of a city (Sassen, 2001, Florida, 2002), the reputation of particular precincts (Cook et al., 2003) and, most importantly, the role of face-to-face communication (Coffey, 1996, Scott, 1999, Pratt, 2000, Sassen, 2000, Sassen, 2001, Coffey and Shearmur, 2002, Pratt, 2002, Cook et al., 2003, Pratt, 2004).

Face-to-face communication facilitates a number of elements in the production of services that cannot be replicated and/or replaced by electronic communication. First, the production of a producer service usually requires a number of inputs from a variety of suppliers. Some of these inputs are structured information that can be freely exchanged and is the same for each user of the information. However, some inputs are dependant on non-standard information or tacit knowledge and cannot be easily communicated in a codified form. For instance a stop sign is a codified message that has one specific meaning and this meaning does not change from one interaction to another. On the other hand, a relatively simple verbal message, such as 'I love you', is a complex, interactive and context dependant message intertwined with the history, trust and understanding of the people communicating, often communicating simultaneously (Leamer and Storper, 2001, Coffey and Shearmur, 2002).

Second, face-to-face communication provides participants with a range of different levels of communication, principally verbal, as well as more subtle communication such as tone of voice, body language and visual cues. These subtle forms of communication enable participants to make judgements about not only an understanding of what is being said but also whether or not to believe it. This element of believability or trust is communicated through these subtle channels of communication (Storper and Venables, 2004, Coffey and Shearmur, 2002) which cannot be replicated by a disembodied voice during a telephone conversation or an email message (Goodchild, 2001, Leamer and Storper, 2001).

Third, face-to-face communication enables simultaneous communication and instant feedback. This enables people to respond to each other, ask for points of clarification and build on a common understanding in an instantaneous way. Storper and Venables (2004) note that simultaneous communication gives participants the flexibility to alter and change their message and/or to add understanding and agreement. This feedback can also contribute to learning on the part of participants. In short, face-to-face communication during the production process can be a rich source of creativity and innovation as clients and suppliers build on each other's ideas (Scott, 1999, Coffey and Shearmur, 2002). While on-line tools such as messenger software and video conferencing provide instant electronic feedback this seems unable to replicate the instantaneous verbal feedback and non verbal cues conveyed during face-to-face contact (Pratt, 2002).

Incorporating face-to-face interaction in producer service production requires a great deal of time which includes planning meetings, preparation, travel time to and from the meeting place, as well as the time spent in a meeting. This time-intensive nature of face-to-face communication encourages producer services firms to locate close to one another in order to minimize the time spent outside of meetings such as that required to travel from the office to the meeting and from meeting to meeting. In addition to reducing the time devoted to formal meetings, spatial proximity enables chance encounters to occur and the informal exchange of information, such as market conditions or employment opportunities, at professional networking functions, over lunch or coffee (Pratt, 2000, Leamer and Storper, 2001, Powell et al., 2002, Pratt, 2002, Pratt, 2004). It is the spatial imperative of face-to-face communication that encourages firms, particularly producer services firms, to agglomerate in specific locations.

Analysis of research into the intra-metropolitan location of producer services confirms this insight into the importance of face-to-face contact. There are two streams. The first examines the location of producer services within a metropolitan area using broad geographic units. The second stream of research uses the actual location of producer services firms within a metropolitan area to identify concentrations of activity.

Examples of the first stream include studies of the intra-metropolitan location of producer services in Sydney (Searle, 1998), Montreal (Coffey and Shearmur, 2002) and Paris (Shearmur and Alvergne, 2002, Guillain et al., 2004). These use data based on predetermined boundaries and use a similar set of industry sectors, notably finance, insurance, real estate, management consulting, market research and accounting. There are some variations in the exact industries studied, however they include broadly similar high order producer services, with the exception of the work on Sydney which also includes graphic designers. These studies show that in Sydney, Paris and Montreal there is a seemingly contradictory spatial pattern with agglomeration in the Central Business District (CBD) on the one hand and dispersion on the other. In Sydney business services are starting to move out of the CBD, with the exception of high order financial services, into the inner suburbs (Searle, 1998). In Paris the finance sector, especially large global banks locate within the CBD while smaller consumer branches and branches designed to service the needs of small businesses are widely distributed across the metropolitan area. Management consultancies displayed a similar pattern with some employment located in the CBD area, particularly large global firms, with many small firms located in the greater Paris metropolitan area (Shearmur and Alvergne, 2002). The research in Montreal examines the location of producer services from 1981 to 1996 and finds a degree of suburbanization of producer services employment although it continued to grow in absolute terms in the CBD. Coffey and Shearmur (2002) suggest that the relative decline of producer services jobs in the CBD is due to increasing specialization of high order producer services, particularly finance and legal services, and the fact that it cannot absorb all the employment growth in producer services. Coffey and Shearmur (2002) also found that the growth in non-CBD producer services employment was concentrated in a small number suburban locations rather than being dispersed across the metropolitan area.

This work provides a useful platform for understanding the intra-metropolitan location of producer services. However, there are a number of shortcomings. First is the use of a mixture of pure producer services, such as management consulting, with mixed producer services, such as finance, that provide both high level services to other businesses as well as basic services to consumers. These have very different markets and inputs and so would be expected to have different locational requirements. Second, administrative boundaries, such as those used for various statistical collections, do not necessarily align with major economic zones. For example, a major industrial area or employment node may be subdivided by administrative boundaries and distort the level of employment concentration. Third, this stream of research has a CBD-centric view of producer services location in that it is seen as the primary location for producer services, but this is not true of all cities or of all producer services.

The second body of work uses the actual geographic location of producer services firms to identify concentrations, free from the distorting effects of pre-determined spatial units. Within this stream of literature work has been completed for London and Vancouver.

In the case of the former Taylor et al. (2003) use the location of firms in financial services, including banking, insurance, accounting, legal service as well as advertising, as the basis for study. In many respects this study updates Goddard's (1975) path breaking research which found that firms were scattered across central London, but there remain distinct and sharply defined geographic agglomerations of firms. Some industries are very highly concentrated with nearly 60 per cent of insurance firms and 55 per cent of banks within these clusters (Taylor et al., 2003).

The actual geography of producer services firms is a more complex matter than a simple dichotomy of the CBD *vs* the rest of the metropolitan area. Rather, there can be sub areas and zones in and near to the CBD that can house these activities of the kind Hutton (2000, 2004a) has illustrated in his research on the location of selected producer services, particularly design orientated industries such as graphic design, multimedia and commercial photographers, in Vancouver. He finds that design intensive producer services in Vancouver concentrate within particular parts of the inner city around the CBD, such as Yaletown, as well as parts of the CBD itself. There is therefore a degree of variety in the location requirements of producer services firms and the CBD is not necessarily the most concentrated, or desired, location for these activities.

The ideas outlined above provide a rich background on producer service location but (apart from some of Searle's (1998) results) they have yet to explore the way in which these location patterns might be evolving over time. We now need to address issues such as: How do location patterns change over time, either as new firms enter the industry, or as new processes or procedures are incorporated into the production process? There are a number of options here. At one extreme we could expect a gradual and spatially incremental dispersal as firms seek out new locations very close to the original core; this implies that any firm in this activity has to be in close to the old agglomeration, and will out-bid present uses to win a location in, or at least on the edge of, the established agglomeration. That outcome is reflected in the replacement of other uses by service sector functions, often seen in new office construction, or the occupation of old industrial space by new services activities. On the other hand we could expect dispersal, either to a new core, or even a series of cores, in an extreme case, to many other sites, as we have seen with manufacturing. That would be reflected in suburban office parks and business estates.

This research therefore explores whether, and how, an agglomeration of a particular service activity that existed in 1981 has changed with the increased use of digital information and the near ubiquitous availability of broadband telecommunication services. Has it dispersed, consolidated or spread incrementally into surrounding areas? The research explores the location of graphic design firms within Melbourne over the period 1981–2001. Graphic design provides an interesting research opportunity as it has often occupied a distinct district within most cities, often in association with media and advertising services. It has also experienced rapid expansion in the number of firms, and also undergone a shift in technology with the introduction of digital technology in the design and printing process. Finally, there is some evidence that parts of the sector can be globalized as the Tombesi et al. (2003) research on architectural drawing services illustrates. Hence we have a sector with an old traditional core district, but many more firms and many new ways of doing business and moving finished product. The central question then is how has the location of firms in this industry changed and what are the drivers?

Method

The method used to identify agglomerations of graphic design firms was to create a geospatial dataset of graphic design firms for 1981, 1986, 1991, 1996 and 2001 identified by using business telephone directories (the Yellow Pages). In order to identify agglomerations of graphic design firms, if they could be identified at all, a spatial statistical program called *CrimeStat* (Levine, 2002) was used. *CrimeStat* uses a nearest neighbor hierarchical clustering algorithm to identify significant groupings or agglomerations of points, in this case agglomerations of graphic design firms.

The nearest neighbor hierarchical clustering routine in *CrimeStat* identifies groups of firms that are spatially closer compared to the mean distance between all firms. This distance is called the threshold distance. The following equation is used to calculate the threshold distance:

$$d(ran) = 0.5\sqrt{\frac{æ A ö}{\c{e} N \o}}$$

where A is the area of the region and N is the total number of firms. The algorithm then compares this threshold distance between firms. Where the distance between two firms is less than this threshold distance the firms are grouped together. The pairs of firms are then compared to another and again where these new pairs are closer together than threshold distance they are grouped together. This process creates larger and larger groupings of firms until either all the firms have been grouped together into one, or a number of agglomerations, or the clustering algorithm eventually breaks down as it cannot agglomerate any more firms together as they are located at a distance greater than the threshold distance. The resulting groups can be displayed as an ellipse using GIS software. *CrimeStat* includes some additional parameters that enable fine tuning of the algorithm. Output is shaped by the values ascribed to key parameters. Extensive sensitivity analysis of *CrimeStat* output was undertaken prior to the choice of values, as discussed in Elliott (2005).

In practical terms the ability to change the parameters creates a subjective element in the identification of agglomerations of firms. One critical decision was to set the number of firms per agglomeration at 15. This was selected in order to identify a few large agglomerations rather than identifying a large number of small agglomerations. As will be displayed below, this technique clearly identifies spatially agglomerations of graphic design firms, and when applied over time, shows the way the location of this activity has evolved.

Establishing the Context: Graphic Design Firms in Australia and Metropolitan Melbourne

Graphic design firms and employment are, as would be expected, a metropolitan activity, and in common with producer services generally, are heavily concentrated in the two largest metropolitan areas of Sydney and Melbourne. In 2001 they accounted for 72 per cent of the 2,400 graphic design firms in Australia. In addition these places also attracted much of the growth in new firms over the past five years (Figure 2.1). That pattern is consistent with the distribution of the people employed in the occupation 'graphic designers' in 2001 (which is only available by State). NSW (Sydney) and Victoria (Melbourne) are the main locations of this type of work and their share of employment is much greater than their share of the total workforce (Figure 2.2).

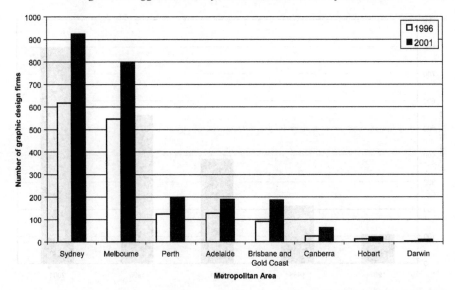

Figure 2.1   Location of graphic design firms by metropolitan area, Australia, 1996 and 2001

Source: Australia on Disk, 1996 and 2001.

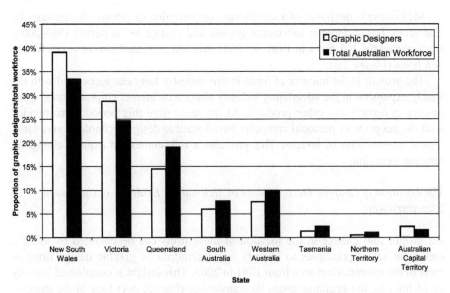

Figure 2.2   Proportion of people describing their occupation as a graphic designer and total population by state, 2001

Source: Australian Bureau of Statistics, Census of Population and Housing, 1996 and 2001.

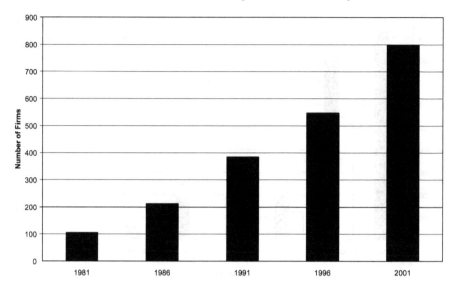

Figure 2.3   Number of graphic design firms in metropolitan Melbourne, 1981 to 2001

Sources: Australia on Disk, 1996 and 2001.
          Telecom Yellow Pages 1981, 1986, 1991.

Melbourne is the home of a significant concentration of graphic design activity. The sector has undergone substantial growth and change in the period 1981-2001; it accounted for 105 firms in 1981; by 2001 this had increased almost seven fold to 798 firms (Figure 2.3).

The growth in the number of firms in this industry has been associated with the steady expansion in the advertising industry along with greater emphasis on graphic images in reports and other products. At the same time this period corresponded with the adoption of personal computer based graphic design technology and, later, digital transmission of images. This provides a rich context to explore change in location over time.

The Location of Graphic Design Firms in Melbourne: Evidence of Continued Concentration?

An insight into change in the location of these firms can be obtained by using a consistent set of principles to identify agglomerations of graphic design firms in parts of the metropolitan area from 1981 to 2001. This insight is established initially by identifying the agglomerations then analyzing changes over time in the share of firms within them, the density of firms within the agglomerations, movement in and out of agglomerations in the past five years, and also exploring the land use categories of land occupied by these firms. This provides a comprehensive set of measures of

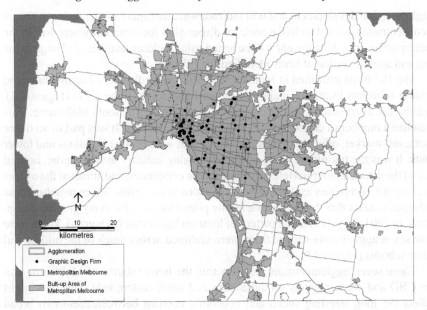

Figure 2.4   Agglomerations of graphic design firms in metropolitan Melbourne, 1981

Source: Telecom Yellow Pages, 1981.

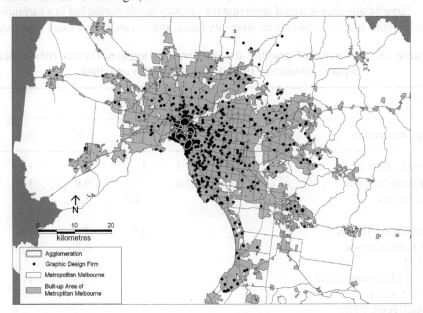

Figure 2.5   Agglomerations of graphic design firms in metropolitan Melbourne, 2001

Source: Australia on Disk, 2001.

agglomeration development and will indicate whether firms have tended to maintain a concentration around an initial node, or dispersal to locations elsewhere within the metropolitan area. That insight will be tested using a closer analysis of change in the original agglomeration of firms identified for 1981.

The 105 firms identified in Melbourne in 1981 were distributed across inner and middle suburban locations, mainly to the east and south of the CBD (Figure 2.4). Within that distribution one agglomeration was identified in South Melbourne. This area had a number of characteristics relevant to this industry. It was part of an office space sub-market, offering CBD access from sites with smaller buildings and lower rents. It was the traditional home of the advertising industry in Melbourne, housed one of the three commercial television and media companies, and attracted the offices of computer companies as that industry expanded in the 1980s. Taken together these influences meant that this area was a highly prized location for many graphic design firms. In 2001, 20 years on, the pattern of location had certainly changed. There were now seven agglomerations, and firms were scattered across much of the middle and outer suburbs (Figure 2.5).

These seven agglomerations are all within the inner suburbs of Melbourne and the CBD and predominantly on the eastern and south eastern side. This bias might reflect the long standing social and economic contrast between these two broad parts of the metropolitan area, where commercial and service businesses and their professional staff, have favoured the east and south, while manufacturing and its lower skilled workers have traditionally located to the west and north.

Superficially then it would seem that the industry has dispersed but in a carefully orchestrated manner, retaining an inner city focus, and in a number of cases retaining

Table 2.1   Density of graphic design firms in agglomerations (firms per square kilometre) 1981, 1986, 1991, 1996 and 2001.

	Year				
Agglomeration	1981	1986	1991	1996	2001
South Melbourne	2.7	7.1	13.1	14.7	14.1
Prahran/St Kilda		3.6	7.5	9.9	14.8
Richmond			4.2	5.5	7.9
Collingwood/Fitzroy			4.6	7.9	12.0
St Kilda/Elwood				6.2	
CBD				13.2	
West Melbourne/CBD					11.1
South Yarra					7.4
South Bank/CBD					7.9

Source: Australia on Disk 1996 and 2001.
            Telecom Yellow Pages 1981, 1986, 1991.

connections or proximity to the original South Melbourne agglomeration. That behaviour seems consistent with the steady rise in the share of all graphic design firms that are in agglomerations. Whereas in 1981 one agglomeration accounted for just 12 per cent of all firms, the seven agglomerations identified in 2001 account for 37 per cent of firms. This suggests that as this industry has grown its firms have looked to locate close by other firms in the industry. That is confirmed by the sharp rise in the density of firms within the agglomerations as seen in Table 2.1. Even in the period 1996–2001 when the share of all firms in agglomerations fell slightly, the density (capturing the tendency of firms to seek out locations close by one another) continued to increase.

Another measure of the change in the significance of location in an agglomeration is the tendency of firms to move within, between, or out of an agglomeration. This analysis involved finding original addresses for firms, and could only be done for sufficient numbers of firms for the 1996-2001 period (Table 2.2). Of the firms that changed location within the agglomeration the majority of these moves were firms that moved within, between or into an agglomeration while 34 per cent of firms moved out of one of the agglomerations. This underlines the value these firms place on a location within an agglomeration.

The significance attached to a location in an agglomeration is confirmed by the short distances that firms have moved either into or out of an agglomeration (Figure 2.7). Over 50 per cent of the firms that have moved out of an agglomeration moved no more than 2 kilometres. This has kept them within easy reach of the locations

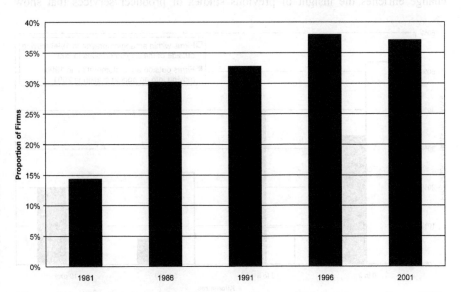

Figure 2.6   Proportion of graphic design firms within agglomerations, 1981, 1986, 1991, 1996 and 2001

Sources: Australia on Disk 1996 and 2001.
          Telecom Yellow Pages 1981, 1986, 1991.

Table 2.2   Movement of established graphic design firms within agglomerations, 1996 to 2001.

Number of firms that moved into the agglomerations	Number of that moved out of the agglomerations	Number of that moved to a different agglomerations	Number of that moved within the same agglomeration	Total
13	21	12	18	64

Source: Australia on Disk, 1996 and 2001.

identified as agglomerations; if a more generous spatial definition of agglomerations had been used it is possible that these relocations would have been contained within the larger boundaries. The in-movers are willing to relocate longer distances: 40 per cent moved between 2 and 5 kilometres, perhaps confirming the premium attached to location within an agglomeration.

This combination of measures suggests that as the location of the graphic design industry has expanded to include the inner city fringe, the CBD has nevertheless remained the key focus accounting for a rising share of firms, accommodated at greater density and also attracting the majority of relocations. This analysis of change enriches the insight of previous studies of producer services that show

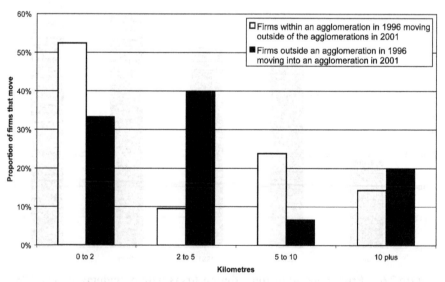

Figure 2.7   Distance of firms moving into or out of one of the agglomerations, 1996 to 2001

Source: Australia on Disk, 1996 and 2001.

strong agglomerations in small parts of cities. It would seem that as the graphic design industry has grown it has retained its concentration, only dispersing within a narrowly circumscribed area.

Firms Outside Agglomerations: Evidence of Dispersal?

It is important to recognize that substantial numbers of graphic design firms operate outside the agglomerations that have been identified. It has been shown (Figure 2.6) that just a third of the total number of graphic design firms in Melbourne were in fact in the agglomerations identified. In addition, more than 60 per cent of the new firms identified on the basis of a comparison of 1996 and 2001 lists of firms have located outside one of the seven agglomerations (Table 2.3). It is possible of course that many of these firms are located 'just outside' the boundaries shown in Figure 2.2. However, it is also apparent that there are many firms distributed across the middle and outer suburbs of Melbourne, which is pointing to dispersal as the common locational response as this industry has grown.

The database used to identify graphic design firms does not provide information about their size (number of employees). However the land use zoning that a firm is located within can be used as a proxy indicator of the size of a firm. Land is zoned for various uses across the metropolitan area; for the purposes of this analysis zoning has been consolidated into five categories:

- Business
- Industrial
- Residential
- Multi-Use
- Other

The purpose was to establish whether firms in agglomerations or outside agglomerations differed in their use of business commercial or residential zones. Although this is a crude indicator it has the potential to discriminate between the type of firms, and perhaps their size, as firms located in residential areas face a number of restrictions on operation and size.

Table 2.3   New Firms identified between 1996 and 2001: location in or outside a agglomeration

	Location in an agglomeration	Locate outside an agglomeration	Total
New firms: number	187	325	512
Share	36.5	63.4	100%

Source: Australia on Disk, 1996 and 2001.

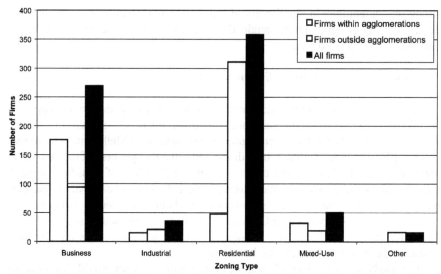

**Figure 2.8   Comparison of land use zoning of firm locations: within agglomera-
tions and outside agglomerations, 2001**

Sources: Australia on Disk, 2001.
          Department of Sustainability and Environment, Planning Scheme Information,
          2003.

Figure 2.8 provides some important information about the graphic design industry in general as well as on the differences in the attributes of firms in agglomerations and outside agglomerations. It shows that a majority of graphic design firms operate from residential addresses; more than 300 of 350 firms in this category are located outside an agglomeration. It is likely that these firms are small, possibly subcontracting to larger organizations, and their location is shaped primarily by the personal circumstances of each firm's principals. In effect, this outcome reflects the availability of lower cost information technology that has allowed small operators to find a niche. In contrast, there are 270 firms located in business zones; they tend to be larger, with more substantial commitment in the form of buildings and capital, and although less numerous probably account for the overwhelming share of employment in graphic design. Significantly, the majority of these firms are in one of the seven agglomerations identified in 2001.

This information places the debate about dispersal and concentration of firms in a new light. It is true that graphic design firms are dispersed across the metropolitan area, and that only one third appear in the spatially constrained agglomerations used here. However, the majority of the dispersed firms are likely to be small, with many operating from residential addresses. Whereas those firms located in business zonings do favour a location in (or near) one of the identified agglomerations, so confirming that concentration at steadily growing density in locations in one sector of the inner city, largely on the CBD fringe, has remained the key focus for the graphic design industries throughout their growth in the 1981–2001 period.

The Expression of Concentration: The Evolution of an Agglomeration of Firms over Time

Table 2.4 is a schematic illustration of the evolution of the seven agglomerations identified in Figure 2.2. It illustrates the fundamental role that the original South Melbourne Agglomeration played in the early development of the graphic design industry. It also shows that new agglomerations have been spawned in what are neighbouring locations to South Melbourne, and not until 1991 was an identifiable agglomeration evident to the north of the CBD (Collingwood/Fitzroy). This confirms that the south-east corridor from the CBD has been the favoured focus of this industry and the number of firms locating there has been expanding and intensifying in density as the industry has expanded. This suggests that the capacity to make and maintain inter firm contact is an important consideration in the location of these firms.

The strength of the concentration forces in this industry can be seen in the simultaneous expansion of firms in the Prahran/St Kilda agglomeration that adjoins the South Melbourne agglomeration. The number of firms in these two agglomerations has essentially proceeded in parallel; only in the 1996 -2001 period has growth in the South Melbourne agglomeration begun to taper off (Table 2.5). This suggests the whole sub-region captured by these two agglomerations expanded as the industry grew. There was not a simple spill over from the initial agglomeration; as new firms arrived they looked for a location in or near the South Melbourne agglomeration, confirming the attractiveness for graphic design services of a small part of the metropolitan area.

Table 2.4   Evolution of agglomerations in metropolitan Melbourne, 1981, 1986, 1991, 1996 and 2001.

Agglomeration	1981	1986	1991	1996	2001
South Melbourne					
Prahran/St Kilda					
Richmond					
Collingwood/Fitzroy					
St Kilda/Elwood					
CBD					
West Melbourne/CBD					
South Yarra					
South Bank/CBD					

Denotes presence of an agglomeration ■

Source: Australia on Disk, 1996 and 2001.
        Telecom Yellow Pages, 1981, 1986, 1991.

Table 2.5   Number of firms within individual agglomerations, 1981, 1986, 1991, 1996 and 2001.

	Year				
Agglomeration	1981	1986	1991	1996	2001
South Melbourne	15	46	53	70	70
Prahran/St Kilda		18	31	44	58
Richmond			13	20	28
Collingwood/Fitzroy			26	35	52
St Kilda/Elwood				16	
CBD				20	
West Melbourne/CBD					26
South Yarra					18
South Bank/CBD					20

Source: Australia on Disk 1996 and 2001
          Telecom Yellow Pages 1981, 1986, 1991

The interdependencies in the changes of the number of firms in these two agglomerations can be observed in the series of maps that shows how they expanded over time (Figures 2.9 to 2.13).

The South Melbourne Agglomeration

The South Melbourne agglomeration is the oldest and largest of the seven agglomerations identified in 2001 and so is the traditional home of the graphic design industry in Melbourne. The South Melbourne area is also the traditional location of advertising agencies especially those located at the top end of St Kilda Road. This illustrates the potential for face-to-face contact between graphic design firms and advertising agencies.

By 1986 the agglomeration became larger in terms of area as well as the number of firms within it, while the number of firms in the Prahran-St Kilda area were sufficiently closely concentrated to justify designation as an agglomeration (Figure 2.10). From 1991 the Richmond area was also identified as a separate agglomeration of graphic design firm. This is the first commercially zoned opportunity to the east of the South Melbourne agglomeration; most of the intervening space is occupied by parkland and sporting facilities. Hence the Richmond agglomeration may be functionally closer to South Melbourne than the distance suggests. Between 1991 and 2001 the shape of the South Melbourne Agglomeration changed very little although it became larger in terms of the number and density of firms. After 1996 the number of firms remained steady, perhaps suggesting this area reached capacity, and

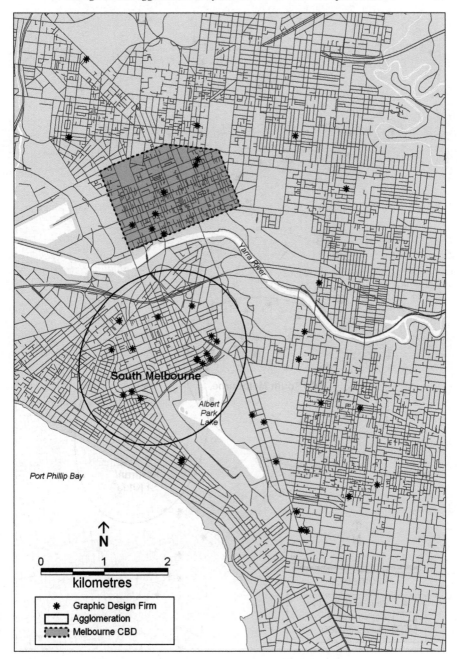

Figure 2.9   Agglomeration of Graphic Design Firms, 1981
Source: Telecom Yellow Pages, 1981.

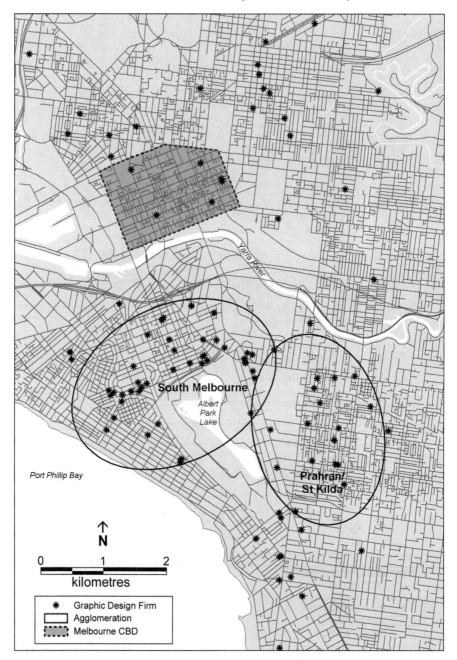

Figure 2.10   Agglomerations of Graphic Design Firms, 1986
Source: Telecom Yellow Pages, 1986.

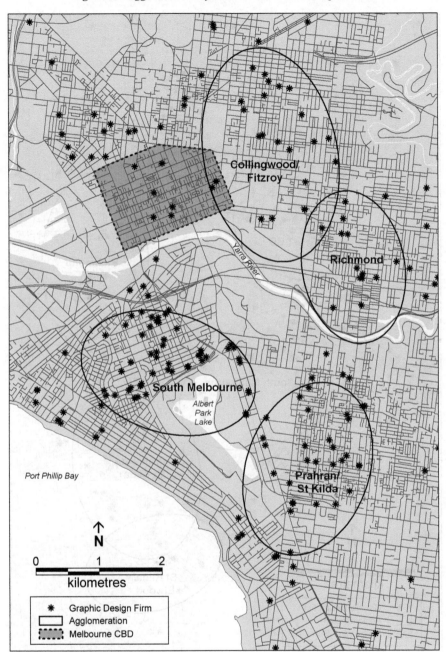

Figure 2.11   Agglomerations of Graphic Design Firms, 1991
Source: Telecom Yellow Pages, 1991.

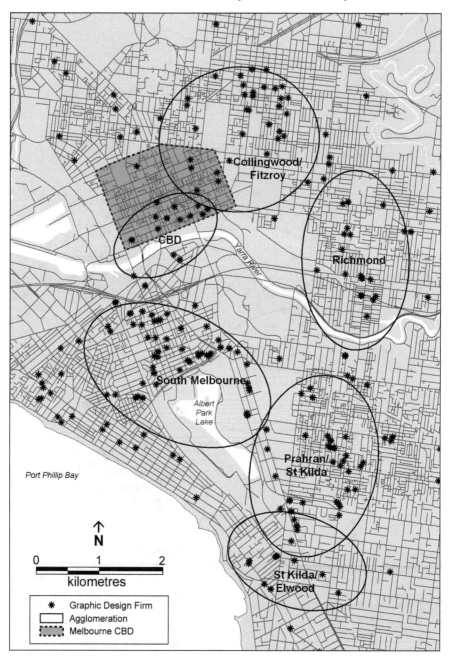

Figure 2.12   Agglomerations of Graphic Design Firms, 1996
Source: Australia on Disk, 1996.

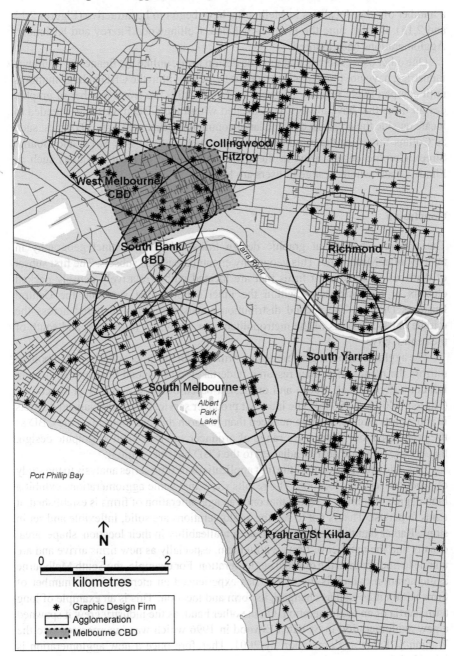

Figure 2.13   Agglomerations of Graphic Design Firms, 2001
Source: Australia on Disk, 2001.

a number of other localities in Melbourne have begun to be attractive (Figures 2.12 and 2.13). By this time the Prahran-St Kilda, Collingwood/Fitzroy and Richmond Agglomerations had enlarged.

One other aspect here is that graphic design firms, unlike the finance or insurance industries, have not found it advantageous to locate within the CBD in large numbers until comparatively recently (Figures 2.12 and 2.13). The small number of graphic design firms that have always been located within the CBD nevertheless tended to locate outside the areas used by the large commercial, financial or head offices, such as Collins Street, or the retail hub in streets such as Swanston Street and Bourke Street. Graphic design firms in the CBD have tended to locate in areas such as Flinders Lane where a range of users are located.

Conclusion

The research shows that graphic design firms exhibit a distinct geography in metropolitan Melbourne. This geography encompasses two subsets. The first subset of distribution consists of firms that are located within relatively densely packed clusters or agglomerations within the inner suburbs that surround the Central Business District. The second distribution consists of firms scattered across the middle and outer parts of the metropolitan area, primarily to the east and south of the central city.

The agglomeration of graphic design firms within the inner suburbs of Melbourne mirrors the results of similar research undertaken on financial business services in London (Taylor et al., 2003) and on design-based business services in Vancouver (Hutton, 2000 and 2004a). The idea that producer services form specialized districts in some parts of cities is more nuanced than research that has focused on the CBD as the prime location. There are a range of producer services, such as graphic design, that choose to locate near or adjacent to the CBD but not within it.

This study provides one of the first applications of time series analysis to the study of producer services location. The results show that some agglomerations exhibit a high degree of inertia. It seems that once an agglomeration of firms is established, it tends to persist. That is not to say the agglomerations are solid, inflexible and set in stone. Rather that they exhibit a degree of malleability in their location, shape, area, density and the number of firms within them, especially as new firms arrive and are unable to find space in the original agglomeration. For example, the South Melbourne agglomeration, which is the oldest, has experienced an increase in the number of firms within it as well as changing its form and location. This is an example of long term inertia of an agglomeration. On the other hand, as the industry grew it spawned a new agglomeration in St Kilda/Elwood in 1996 which was incorporated into the Prahran/St Kilda agglomeration by 2001. Therefore once a new agglomeration is identified there is no certainty that it will retain its initial configuration as new and established firms move around seeking to find locations that suit their operations.

Changes in the location of producer services are not unexpected. The location of financial services in New York have been changing over time as the centre of activity has shifted from Wall Street to other locations such as Midtown. The reasons for this change are a combination of technological change, such as communications technology, organizational change, such as the splitting of front office and back office functions and, to a lesser extent, the result of the terrorist attack on September 11 (Pohl, 2004). Yet the old core of lower Manhattan, like the core of graphic design in South Melbourne, remains a significant location for this industry.

The results presented here also go to the heart of a debate about the impact of telecommunications and other media on firms producing 'weightless products', transported anywhere with the use of information and communication technologies. Although it is imaginative to suggest that this new context could enable firms to locate in any part of the metropolitan area and avoid spatially related costs and limitations such as high commercial rents in the inner city or difficulties of parking in the inner suburbs, that is not consistent with the findings here. Rather it seems that some special features of the inner city interact with the needs of firms and produce agglomerations of firms; even as the industry grows the spatial spread is often to adjoining locations. In short it would seem that reports of the death of geography are greatly exaggerated.

This research also contributes to the wider idea of the remaking of the economic geography of the city and the renewal of the inner city. The renewal and remaking of local economic geography is a phenomenon that is occurring in most inner areas of large areas. London (Taylor et al., 2003), Vancouver (Hutton, 2004a and 2004b) and Boston (Glaeser, 2005) have received recent attention. In the same vein, Inner Melbourne witnessed a sea change in its spatial economy. In the 1950s it was the location for head offices of resource companies, banks, retailers and manufacturers. However there have been some marked changes in the location of some of these industries. For instance manufacturing in the inner city has declined significantly from employing 143,000 people (60 per cent of manufacturing workers in the metropolitan area in 1949 (MMBW, 1954)) to approximately 30,000, or 12 per cent of manufacturing workers in the metropolitan area in 2001 (ABS Census of Population and Housing, 2001). Graphic design firms have been prominent among the new firms occupying the space left vacant as a result of this change, along with activities such as management consultants and parts of the computer industry. This research has demonstrated the spatially selective and uneven nature of this renewal. Areas wax, such as South Melbourne and Collingwood, while others wane, such as Richmond. Some areas, for the time being at least, miss out entirely, such as Footscray to the west or Brunswick to the north of the CBD, as the economic landscape of the inner city is remade. An understanding of producer service location remains a powerful mechanism for understanding the re-shaping of city prospects.

A significant contribution of this research is its methodology. By connecting a readily available source of data to an established GIS-based data manipulation routine it has provided a new way of exploring the locational patterns of producer services in cities. Analysis of the latter has often been constrained by data availability, as employment by occupation or industry has not been available in the small spatial units that would allow detailed intra-urban analysis. Where they are available, the source is often national censuses, which in some countries can mean ten year gaps in data supply. Reports by commercial property firms provide a more frequent but less comprehensive and objective insight on the way office space markets (and hence opportunities for producer service firms) are working. Faced with this circumstance much producer service research has relied on surveys, which provide rich and detailed insight on some firms in some areas, but can lack a national or metropolitan-wide perspective and so limit the contribution to understanding the sector overall.

Published telephone directories, or other commercial guides to selected sectors, provide the possibility of regular monitoring of patterns of firm location and (as shown in this chapter) can expose major shifts in the location of activity over time. Although the ready availability of the data makes it attractive it obviously has some drawbacks: it does not differentiate by size of firm, for example, which is a very serious issue. That was addressed in the current research by the check on land zoning associated with each location, but that is an awkward step to take. So directory data with addresses does provide a new way to look at producer service firms, but it carries some limitations.

The major innovation in this chapter, and the greatest potential for further research, is the use of a GIS package to visually represent locations. Visual output showing firm location has a long heritage in economic geography and in producer services research, but was generally limited to output from surveys. The approach developed here provides a metropolitan area perspective, with the capacity for closer study of small areas. It can also draw in other information in overlays: the location of firm that are expected to have connections and linkages with, for example, local infrastructure and so on. Finally it provides the chance to look at change over time in a matter of seconds. Although this visual capacity does not 'solve' any issues, it sharpens the questions that we can ask about this sector and can provide insight on many aspects at a speed and comprehensiveness that was not available earlier. The analytical edge created by the clustering algorithm opens up some new approaches to some old issues. First it is possible to explore different types of definitions of the nearest neighbour concept and find tight and loose definitions of an agglomeration: the geographic differences provide insight on the way a core area and a periphery area of these services emerges. The approach could also compare the overlap of agglomerations of different services and how that has changed over time.

Hence the technique used here provides great scope to explore the way in which producer services locate in cities. However, although it has the advantages of easy data access, comprehensiveness and speed it can never really 'explain' the locational behaviour of firms. There the interview will remain the central methodology. The approach used here will help to find firms to interview, and also help to shape the questions to ask them. As such it is a very valuable step forward in research on producer services.

References

Australian Bureau of Statistics (2003), Census of Population and Housing 2001, Canberra: ABS.

Bathelt, H. (2002), 'The Re-emergence of a Media Cluster in Leipzig', *European Planning Studies*, **12**, 583612.

Bowen, J., Leinbach, T.R. and Mabazza, T. (2002), 'Air Cargo Services, the State and Industrialisation Strategies in the Philippines', *Regional Studies*, **36**, 451–467.

Cairncross, F. (1997), *The Death of Distance: How the Communications Revolution will Change our Lives*, Boston: Harvard Business School Press.

Coffey, W.J. (1996), 'Forward and Backward Linkages of Producer-Service Establishments: Evidence from the Montreal Metropolitan Area', *Urban Geography*, **17**, 604-632.

Coffey, W.J. (2000), 'The Geographies of Producer Services', *Urban Geography*, **21**, 170-183.

Coffey, W., Drolet, R. and Polese, M. (1996), 'The Intra metropolitan Location of High Order Services: Patterns, Factors and Mobility in Montreal', *Papers in Regional Science*, **75**, 293-323.

Coffey, W.J. and Shearmur, R.G. (2002), 'Agglomeration and Dispersion of High-Order Service Employment in the Montreal Metropolitan Region, 1981–1996', *Urban Studies*, **39**, 359-378.

Cook, G.A.S., Pandit, N.R., Beaverstock, J.V., Taylor, P.J. and Pain, K. (2003), 'The Clustering of Financial Services in London', GaWC Research Bulletin 124, Loughborough University.

Crevoisier, O., and Maillat, D. (1991), 'Milieu, Industrial organisation and territoprial production System: towards a new theory of Spatial development', in Camagni, R (ed.) *Innovation Networks. Spatial Perspectives*, London: Belhaven for GREMI., pp. 12–34.

Department of Sustainability and Environment (2004), *Urban Development Program Report – 2004*, Melbourne: State Government of Victoria.

Dicken, P. (1998), *Global Shift: Transforming the World Economy*, London: Paul Chapman Publishing Ltd.

Elliott, P.V. (2005), 'Intra-Metropolitan Agglomerations of Producer Services Firms: The Case of Graphic Design Firms in Metropolitan Melbourne, 1981 to 2001', unpublished Master of Planning and Design Thesis, The University of Melbourne.

Florida, R. (2002), *The Rise of the Creative Class*, New York: Basic Books.

Glaeser, E.L. (2005), 'Reinventing Boston, 1630–2003', *Journal of Economic Geography*, **5**, 119–153.

Goddard, J. (1975), *Office Location in Urban and Regional Development*, London: Oxford University Press.

Goodchild, M.F. (2001), 'Towards a Location Theory of Distributed Computing and E-Commerce', in Leinbach, T.R. and Brunn, S.D. (eds), *Worlds of E-Commerce: Economic, Geographical and Social Dimensions*, New York: John Wiley and

Sons, pp. 67–86.

Graham, S. and Marvin, S. (2001), *Splintering Urbanism: Networked Infrastructure, Technological Mobilities and the Urban Condition*, London: Routledge.

Guillain, R., Le Gallo, J. and Boiteux-Orain, C. (2004), 'The evolution of the spatial and sectorial patterns in Ile-d-France over 1978-1997', available at, http://139.124.177.94/proxim/viewpaper.php?id=245&print=1, Retrieved 5/5/2005.

Guiliano, G., Redfearn C., with Agarwal, A., Li, C and Zhuang, D. (2005), 'Not all Sprawl: Evolution of Employment Concentrations in Los Angeles, 1980 – 2000', School of Policy, Planning and Development, University of Southern California, Los Angeles, California, USA. Paper submitted for publication in *Urban Studies*.

Hutton, T.A. (2000), 'Reconstructed Production Landscapes in the Postmodern City: Applied Design and Creative Services in the Metropolitan Core', *Urban Geography*, **21**, 285–317.

Hutton, T.A. (2004a), 'The new economy of the inner city', *Cities*, **21**, 89–108.

Hutton, T.A. (2004b), 'Post-industrialism, post-modernism and the reproduction of Vancouver's central area: Retheorising the 21st-century city', *Urban Studies*, **41**, 1953–1982.

Lambooy, J.G. (1997), 'Knowledge Production, Organisation and Agglomeration Economies', *GeoJournal,* **41**, 293–300.

Leamer, E.E. and Storper, M. (2001), 'The Economic Geography of the Internet', Working Paper 8450, National Bureau of Economic Research, Cambridge, Massachusetts.

Levine, N. (2002), *CrimeStat 2.0: A Spatial Statistical Program for the Analysis of Crime Incident Locations*, Ned Levine and Associates and the National Institute of Justice.

Marshall, A. (1920), *Principles of Economics*, London: Macmillan and Co., Ltd, 8th edition.

Melbourne and Metropolitan Board of Works (1954), *Melbourne Metropolitan Planning Scheme: Survey and Analysis*, Melbourne: McLaren & Co.

Moss, M.L. and Brion, J.G. (1991), 'Foreign Banks, Telecommunications and the Central City', in Daniels, P. (ed.) *Services and Metropolitan Development: International Perspectives.* London: Routledge, pp. 265–284.

O'Brien, R. (1992), *Global Financial Integration: The End of Geography*, London: The Royal Institute of International Affairs.

O'Connor, K. (1991), 'Creativity and Metropolitan Development: A Study of Media and Advertising in Australia', *Australian Journal of Regional Studies*, **6**, 1–14.

Piore, M.J. and Sabel, C.F. (1984), *The second industrial divide: possibilities of prosperity*, New York, Basic Books.

Pohl, N. (2004), 'Where is Wall Street? Financial Geography After 9/11', *The Industrial Geographer*, **2**, 72–93.

Powell, W.W., Koput, K.W., Bowie, J.I. and Smith-Doerr L. (2002), 'The Spatial Clustering of Science and Capital: Accounting for Biotech Firm-Venture Capital

Relationships', *Regional Studies*, **36**, 291–305.

Pratt, A.C. (2000), 'New Media, the New Economy and New Spaces', *Geoforum*, **31**, 425-436.

Pratt, A.C. (2002), 'Hot Jobs in Cool Places. The Material Cultures of New Media Product Spaces: The Case of South of the Market, San Francisco', *Information, Communication and Society*, **5**, 27–50.

Pratt, A.C. (2004), 'The Cultural Economy: A Call for Spatialized 'Production of Culture' Prespectives', *International Journal of Cultural Studies*, **7**, 117–128.

Pryke, M. (1991), 'An International City Going Global: Spatial Change in the City of London', *Environment and Planning D: Society and Space*, **9**, 197–222.

Quah, D.T. (1996), 'The Invisible Hand of the Weightless Economy', Occasional Paper 12, Centre for Economic Performance, London School of Economics.

Saxenian, A, (1990), Regional networks and the resurgence of Silicon Valley, *California Management Review*, **33**, 89–112.

Sassen, S. (2000), *Cities in a World Economy*. Thousand Oaks: Pine Forge Press.

Sassen, S. (2001), *The Global City*, Princeton: Princeton University Press.

Scott, A.J. (1988), 'Flexible Production Systems and Regional Development: The Rise of New Industrial Spaces in North America and Western Europe', *International Journal of Urban and Regional Research*, **12**, 171–186.

Scott, A.J. (1999), 'The Cultural Economy: Geography and the Creative Field', *Media, Culture and Society*, **21**, 807–817.

Scott, A.J. (2000), 'The Cultural Economy of Paris', *International Journal of Urban and Regional Research*, **24**, 567–582.

Scott, A.J. and Storper, M. (2003), 'Regions, Globalization, Development', *Regional Studies*, **37**, 579–593.

Searle, G.H. (1998), 'Changes in Producer Services Location, Sydney: Globalisation, Technology and Labour', *Asia Pacific View Point*, **39**, 237–255.

Shearmur, R. and Alvergne, C. (2002), 'Intra-metropolitan Patterns of High-Order Business Service Location: A Comparative Study of Seventeen Sectors in Ile-de-France', *Urban Studies*, **39**, 1143–1163.

Storper, M. and Venables, A.J. (2004), 'Buzz: Face-To-Face Contact and the Urban Economy', *Journal of Economic Geography*, **4**, 351–370.

Taylor, P., Beaverstock, J.V., Cook, G., Pandit, N. and Pain, K. (2003), *Financial Services Clustering and its Significance for London*, London: The Corporation of London.

Thrift, N. (1994), 'On the Social and Cultural Determinants of International Financial Centres: the Case of the City of London', in Corbridge, S., Martin, R. and Thrift, N. (eds), *Money, Power and Place*, Oxford: Blackwell, pp. 327–355.

Tombesi, P., Dave B. and Sriver, P. (2003), 'Routine Production or Symbolic Analysis? India and the Globalisation of Architectural Services', *The Journal of Architecture*, **8**, 63–94.

Wise, M.J. (1949), 'On the Evolution of the Jewellery and Gun Quarters in Birmingham', *Transactions of the Institute of British Geographers*, **15**, 57–72.

Urry, J. (1990), "On the Social and Cultural Determinants of International tourism", in Thrift, N. and Williams, P. (eds), *Class and Space: The Making of Urban Society*, Routledge, London.

Chapter 3

Development of the 'Third Form' of the Car-making Producer Services Industry in Shanghai, China and its Locational Factors

Yufang Shen[1]

Introduction

The discussion of the producer services industry has become a hot topic during recent years in the circle of the so-called 'new economic geography' along with the fast growth of the sector and, notably, it increasingly becomes the prime source of sustained high value added as Gibbons et al. mentioned (Gibbons et al., 1994).[2] It, to certain extent, represents a new era (or an endorsement) of the geographers in conceiving the dynamics of economic change and new geography. In this, many efforts have been made, e.g. Daniels (1991, 1993) from an international angle explored the inter-relationship between growth of the sector and the development of metropolitan areas, and its characteristics and development tendency in the world economy; Daniels et al. (1993) inquired into the geography of services; Marshall and Wood (1995) probed into key aspects of urban and regional development via searching links between services and space; Bryson, Daniels and Warf (2004) discussed people, organizations and technologies in service worlds; Daniels, Hutton and Ho (2005) studied the interrelated complexes of the service industry with other economic activities and cities, and development trajectories in the Asia Pacific; and so on. More specifically, Daniels and Moulaert (1991) had made a thorough inquiry into the changing geography of advanced producer services; Arguelles and Morollon (2004) approached factors affecting outsourcing decisions in advanced business

1 Member of the Standing Committee of China's National Society on Economic Geography (The Yangtze Basin Development Institute, East China Normal University, Shanghai 200062).

2 Re-cited from Camacho, J.A. and Rodriguez, M. (2004), 'Embodied knowledge flows and services: an analysis for six European countries', paper presented at Annual Conference of the IGU Commission on Dynamics of Economic Space, Birmingham, UK, 10–13 August, 2004.

services using discrete response models when studying the outsourcing patterns of advanced business services in the Spanish economy.

Moreover, the third form of the car-making producer services industry refers to a totally new generation of service economy in the world today. Although the notion has not yet been clearly defined in the literature, by extension and change of the relevant term of 'producer services' in the *Dictionary of Human Geography* (see Thrift, 2002), we can still find some clues to the nature and characteristics of this particular kind of industry. And, as such, it can then be defined as an industry that supplies car-related services to businesses and governments including directly to individual consumers. Such industry is characterized as that which provides 'intermediate' inputs into both the process of production and sales to end consumers (e.g. cultural entertainments and sports) and as entirely different from the ones that existed before. Here, the fact is that it can provide many car-related and value-added businesses – through concentrated presentation of the brands and after-sale service, extension of relative R&D activities and related cultural entertainments and advertisement besides car-marketing – all on the back of the emergence of 'new economic spaces' and new geography. Contemporarily, it is one of the high points in a knowledge-based economy worldwide, representing a significant trend of the future economy of the twenty-first century.

In the literature, however, the discussion of the nature, impacts and specifically locational factors of this sector is lacking. Thus, for this reason, to probe into its nature, characteristics, effects and thereby locational factors is undoubtedly helpful in terms of either promoting further development of the new geography literature, or lifting up the level of the regional economies and their competitiveness practically.

The question arising here is that, despite the above apparent abundance of analyses, a detailed understanding of the industry itself and its relevance to the dynamics of the new economy and new geography is still weak. Thus far, there has been little systematic study concerning the nature and impacts of the rise of the industry, or the like, in a regional context. This has inevitably affected our better understanding of the changes and development in service worlds. Whilst many studies generally agree that the changes in the global economy and the restructuring process in firms are the main reason for industrial changes in service worlds, there is little analysis as to what extent these changes affect China's local economy and how far such changes may need to be re-justified with respect to creating a new economy locally on the back of the globalization process. Within this context, this chapter argues that the 'third form' of the industry recognizes the interconnections among marketing, design, leasing, car-related R&D, entertainments and financial services and that development of the industry certainly links to the rise of the new economy and, also, considers its locational determinants as the main components in promoting the growth of the local economy on the one hand and the dynamics of new economic spaces on the other.

Accordingly, and methodologically, based on observation this chapter uses positive and inductive probation approaches and lays its specific emphasis on the issue of locational determinants. The case study was of Shanghai International Car

City and most of the main evidence and relevant data come from field investigation of the City, local authority reports and government's websites. This involved four interviews with the selling departments of the Shanghai Volkswagen Co., Ltd. and the Shanghai International Car City Development Co., Ltd. carried out by my students and secondary data from reports of the Foreign Economic Commission, Anting District in Shanghai and websites of the Anting District's local government and the Shanghai International Car City. Among these, data from the local authorities' reports and websites are the most important and useful source in the justification of the main findings and, in particular, locational determinants.

About the 'Third Form' of the Car-making Producer Services Industry

Its Nature and Difference

The development of the car-making producer services industry has passed through roughly three stages in history. The generation of the first stage (or, the first form) of the car-making producer services industry dates back to the middle of the1990s (and even before) and, by and large, it was mainly concerned with the car-selling business only (Zhang and Wu 2003, p. 17).

Since the end of the 1990s, moreover, with many exclusive car-dealing stores named '4S'stores (an integrated service model of sales, spare parts, service and survey), the so-called 'second stage' (or, the 'second form') of the car-making producer services industry has emerged, which has lasted up until now (same as above).

In this, the 'third form' of the car-making producer services industry is an entirely new model and represents the process of the third stage in this regard. Comparatively, the main difference between the third and the other two forms of the car-making producer services industry is that it combines with more functions than the other two, including not only car-marketing, car-related logistics, R&D, commercial, financial, lease, accounting, insurance and legal services, but also car-related cultural salon, entertainments and sports, food and shopping, etc.

At present, although it is not yet clear when this kind of economy will dominate the business, it indicates a surely inevitable trend of future development of the new generation of service economy in general and the economy of the car-making producer services industry in particular. For Shanghai, the building of the super-class car market, Shanghai International Car City, with multi-functions of modern services will certainly represent the latest level of development of the new generation of the car relevant economy in China.

Its Chinese Background

The 'third form' of the car-making producer services industry in Shanghai has been generated on the backdrop of the globalization process of the world economy and China's accession to the WTO. In this, it should be mentioned that in Chinese consensus, because the car-making industry in China is still immature, it will probably be weakened thereafter with the loss of tariff protection. Due to tariff concession and quota cancellation after China's accession to the WTO, China's car-making industry will be faced with huge pressure and, therefore, a big problem will be how to survive when cars imported from western countries dominate the Chinese car market. Consequently, in order to meet the challenges of the contemporary world, it is very important for China to develop its car industry by full use of resources from both Chinese and foreign transnational corporations and by the development of the new generation of the car-making producer services industry. For the above reason, the building of the Shanghai International Car City has been, and can be, seen as a good attempt in resolving this kind of problem and dilemma.

Development of the Shanghai International Car City and its Main Functions and Features

Since mid-2003, along with the boom of the car-selling market, a big change has taken place in the car-making producer services industry in Shanghai, the largest and one of the four centrally administrated municipalities in China situated in the middle of the eastern coast and alongside the outlet of the Yangtze River (see Figure 3.1).

This sees the rapid growth of a new generation of the car-trading market and market place which is different from ever before and has been called 'the third form' of the car-making producer services industry with the building of two giant car-trading service cities plus three car-trading service avenues are on the way within the central area of Shanghai. Among them, the Shanghai International Car City is the largest. With some billion yuan of input, this Car City includes almost all modern businesses of the car-making producer services industry, e.g. car-marketing, car-related logistics, R&D, commercial, financial, lease, accounting, insurance and legal services, car-related cultural salon and entertainments, food and shopping and, to a certain extent, represents the trend in this regard.

By plan, the building of the Shanghai International Car City with six integrated functions in car trading, car-related exhibition, R&D, logistics, tourism and other relevant services will be completed at the Anting Town, Jiading District in Shanghai during the Tenth Five-Year Plan period between 2001 and 2005 (see Figure 3.2). Therefore, it will become China's key market and one of the important distribution centres and providers in the Asia Pacific Ring for car-trading and other modern services, symbolizing the formal rise of the 'third form' of the car-making producer services industry in China.

Its Locational Structure

Accordingly, and geographically, the Shanghai International Car City includes areas of car racing, spare parts making, car manufacturing, core marketing and

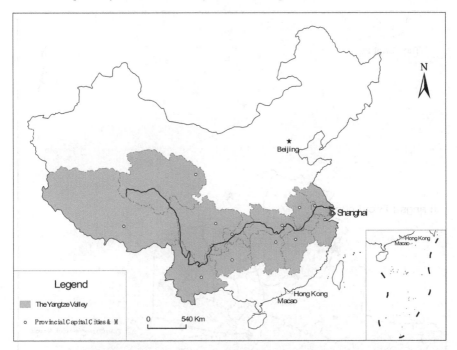

Figure 3.1   Map of administrative areas of the Yangtze Valley and the location of Shanghai

R&D activities, an educational garden, a theme park, old and new towns and a golf course. In sum, it comprises in total five main sub-divided functional areas detailed as follows.

The Educational Garden The Educational Garden in the Shanghai International Car City covers a total land space of 1.7 square kilometres. In order to maintain the necessary manpower resources for the Car City, some of the leading universities will be attracted into and be settled in this Garden.

The Anting New Town Area The Anting New Town Area covers in total 5.4 square kilometres of land space and is actually a packaged residential area of the Shanghai International Car City.

Manufacturing Area of Car-assembly and Spare parts Production The manufacturing area, covering a land space of 15.5 square kilometres, is one of the most important car manufacturing bases in China specializing in car assembly and spare parts production. In this area, there exist car-making factories of the Shanghai Volkswagen Company and the Shanghai Automotive Industry Corporation Group (SAIC) as well as the Shanghai International Car-parts Zone covering in total 8.3 square kilometres of land space.

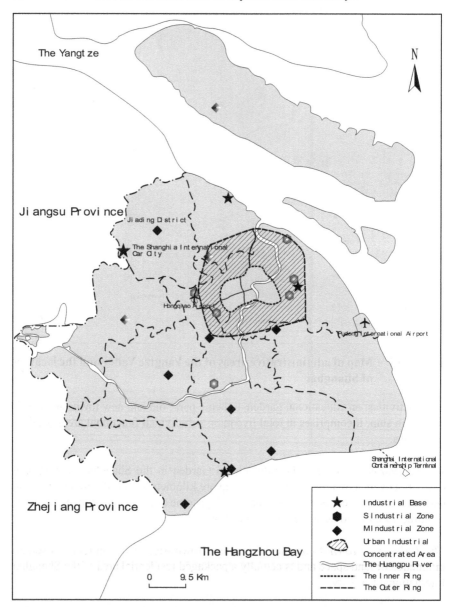

The Yangt ze

Ji angsu Pr ovi nce

Ji adi ng D st ri ct

The Shanghi a I nt er nat i onal
Car Ci t y

Hongqi ao A rpor t

Pudong I nt er nat i onal Ai r por t

Zhej i ang Pr ovi nce

The Hangzhou Bay

Shanghai I nt er nat i onal
Cont ai ner shi p Ter mi nal

0   9. 5 Km

★	I ndust ri al Base
⬢	S I ndust ri al Zone
◆	M I ndust ri al Zone
⬭	Ur ban I ndust ri al Concent r at ed Ar ea
～	The Huangpu Ri ver
----------	The I nner Ri ng
------	The Out er Ri ng

Figure 3.2   The location of the Shanghi International Car City

The International Car-racing Field The International Car-racing Field is located
in the northeast of the Car City covering a land space of 5.3 square kilometres. It
contains a car-racing field of the Shanghai International Circuit, a commercial and
exhibition area, a cultural and entertainments area and a reserved area for future
development. The total investment in this field amounts to 5.6 billion yuan. From

24–26 Sept 2004, the FIA Formula One World Championship Chinese Grand Prix will be held at Shanghai International Circuit. The season culminated in 2004 with three fly-away races, the first of which was the Grand Prix of China in Shanghai on September 26.[3]

The Core Area   The Core Area covers 7.4 square kilometres of land space in total. By plan, within it there will be an exhibition and commercial area, an R&D zone, a theme park and a golf course. At present, most of the construction projects in this area have been completed to plan and it will become an integrated multifunctional area for car exhibitions and commerce, car-related R&D activities, information service, scientific education, culture and entertainment, tourism, logistics and storage.

Its Main Functions

Car Trading   This is the main function of the Car City. Being backed by the advantages of the sea port, industries, market potential, relatively advanced financial service and international influence of Shanghai and supported by the municipality, the Car City will use the policy tool to attract both domestic and international major car-trading companies to it, making it an important platform for cooperation and competition between Chinese and international car makers. In this, to attract more marketing headquarters or sub-headquarters of the internationally famous car makers to set up their businesses in the Car City is the focus. The context of the car-trading function includes: i) sale of cars; ii) trade of spare parts and accessories; iii) second hand car trading; and iv) e-commerce.

Car-related Scientific Education and Exhibition   Using the concentrative exhibition effect of the internationally famous car brands, the Car City will allow free information exchange with the outside world to direct and guide the trend of China's car-making and consumption businesses and, at the same time, to popularize the knowledge of car-related science, technology and culture to improve scientific and cultural qualities and the predisposition of the public, the car consumption groups and the market. The context of this function is concerned with: i) presentation of the brands; ii) international fair; iii) car exposition and commerce; iv) car-related education and training; and v) scientific exchange.

Car-related Services   The car-related service function is the basic foundation for all other functions. In this, the Car City will certainly, and determinedly, develop its modern car-related service industry as its main business to change China's existing models in car dealing, car-related consumption credit and insurance following world practice and thereby to promote a healthy development of the Chinese car consumption market. The context of car-related service function contains: i) one-

---

3   Data from http://www.icsh.sh.cn, 22 April, 2004.

stop service in car-trading, ii) car-related finance and information; iii) car-related club business; iv) car lease; and iv) car-related maintenance.

Car-related Culture and Tourism   The function of car-related culture and tourism is an effective instrumentality in building up the Car City's overall image, creating car-related cultural milieu, lifting up the car-related cultural qualities of the consumers and in delivering car-related cultural values. It will then inevitably become a hot tourist spot with new and beautiful scenery and peculiar identity to Shanghai. The context of car-related cultural and tourist function includes: i) popularization of car-related science and technology; ii) car-related recreation and entertainments; iii) car-related sports; and iv) car-related tourism.

Car-related Logistics   The car-related logistics function is one of the important characteristics of the Car City in terms of modernization and internationalization. It is a bridge that can link car makers, suppliers, distributors and consumers and interlock them in one place and, thereby, guarantees perfecting of other relevant functions of the Car City. For this reason, using the seaport advantages of Shanghai and based on the Waigaoqiao Free-trade Zone, the Car City is now trying its best to build up a relatively advanced logistics network system so as to attain the objectives of efficiency and lower cost. The context of car-related logistic functions includes: i) storage and transportation; ii) packing and just-in-time delivery service; and iii) optimization of the supply chains.

Car-related R&D and Manufacturing Activities   The car-related R&D and manufacturing activity function is an important domain in supporting development of relevant industries in the Car City. In this, based on existing industries and further development, it will therefore accelerate its R&D and manufacturing activities for new products and technology so as to maintain its international competitiveness in cost and quality. The context of car-related R&D and manufacturing activity function comprises: i) research and development of new products, models and state-of-the-art technology. In this, there will be a car-making technology R&D centre, an information centre, a supply technology R&D centre and centers for endorsement, education and training in the Car City; and ii) car-assembling and spare parts production. By now, besides the Shanghai Volkswagen Company, there have been over 100 enterprises of the TNCs from Germany, the United States, France, Japan, Korea, Singapore and Taiwan settled in this city. Along with the further development of the Car City, there will be more internationally leading car manufacturing and spare parts-making companies move in.

Its Main Features

Entirely Internationalized   The Shanghai International Car City is envisaged, and has been constructed, to be one of the most important international distribution hubs for the trading of cars and materials and the exchange of information in both the

Chinese domestic and Asian markets. In the forthcoming two-three years, it will try to encourage the three contemporary and internationally leading car-making manufacturers – General Motor and Ford of the United States, Volkswagen of Europe and Toyota and Honda of Japan – to locate their trading headquarters in the Car City.

Large-scale of Economies By plan, in the next five years, the Shanghai International Car City will try to maintain its car and spare parts trading share of no less than 15-25 per cent of that of the Chinese market as a whole. Moreover, after tariff exemption and abandonment of the quota protection by the WTO regulation, the share of China's total market is expected to exceed 30-40 per cent. In 2010, the total amount of sales via the Car City is expected to reach 150-250 billion yuan.

Relatively Advanced By plan again, encouragingly and ambitiously, the Shanghai International Car City will attract and use the latest achievements in car science and technology and, meanwhile, introduce the latest sense and ideas in car-dealing business following the WTO regulation and international practice. It will then become an international distribution centre with the characteristics of openness and regionalism in car-dealing and spare parts-trading.

With Comprehensive Functions Relying and resting on Jiading District's existing car-making establishments, the Shanghai International Car City will speed up its car exhibition and commercial function by playing both the 'China Card' and the 'International Card'. It will then become an integrated, modern, comprehensive and internationalized car city with multi-functions.

With All-round Car-related Services The Shanghai International Car City will attract all necessary corresponding institutions of the industrial and commercial bureau, tax office, bank, insurance company and car-control office to set up their business in the city and work together with them in an all-in-readiness mode so as to provide a 'one-stop' service to both domestic and foreign customers. These will certainly be helpful in establishing an international image for the Shanghai International Car City with high-efficiency and convenience.

Influence of the 'Third Form' of the Car-making Producer Services Industry on the Overall Growth of the Car-making Industry in Shanghai

Promotion of the Car-making Industrial Cluster

Generally, a car-making industrial cluster indicates many different kinds of car-making industrial establishments set in a place (or an area) that is fit for making cars with a concentration of necessary transportation facility, spare parts production, information, R&D and many other relevant service (or supporting) business and thereby forms a car-making industrial base.

The experience from developed countries tells us that it is an inevitable trend to form the car-making industrial cluster in the process of development of the car-making industry in the contemporary world. For example, in Japan there existed a car parts manufacturing factory in Yokohama hundreds of kilometres away from the headquarters of the Toyota Corp. This seriously hindered the regular production performance of the Toyota Corp. In order to correct this, the factory has now been moved closer to the headquarters. Meaningfully, the car-making industrial cluster in Detroit in the United States has also formed spontaneously.

Undoubtedly, putting huge sums of investment into the development of the Car City in Shanghai has been the municipal government's focal point so as to speed up development of the local car-making industry on one hand and to form a car-making industrial cluster in Shanghai, or more broadly, in the Yangtze Delta Area on the other. Consequently, it has been envisaged that it can increase the international competitiveness of China's car-making industry shortly and greatly.

Positive Effects on Other Aspects of the Local Economy

By using experience from western countries, again, development of the 'third form' of the car-making producer services industry in general, and the Car City in Shanghai in particular, will certainly exert positive effects on other aspects of the local economy. These include the contributions from the industry to state tax, revenue and employment, etc.

Tax, Revenue and Balance of the Foreign Exchange   Statistics have estimated that the annual output value of the car-making industry in the world as a whole amounts to about 1,500 billion US dollars, with that of the United States at 400 billion, Germany, France, UK and Italy at 400 billion, Japan at 320 billion and Korea at 50 billion. In 1997, the proportion of the gross output value of the car industry in the United States, Japan, Germany, France, Korea and Spain in total GDP were 4.2 per cent, 7.2 per cent, 10.5 per cent, 5.4 per cent, 12.6 per cent and 9.2 per cent respectively and the proportion of the net increase output value of the car industry of the above corresponding countries in total GDP were 1.42 per cent, 2.1 per cent, 3 per cent, 1.6 per cent, 3.4 per cent and 2.6 per cent respectively (Industrial Development Bureau of the United Nations, 1997).

The rapid growth of the economy of Germany in the 1950s and the taking-off of the economy of Japan in the 1960s showed that there were somewhat the same inherent relationships between rapid growth of the economy and the high-speed development of the car industry. In this, the contributions of the car industry to state tax, revenue and employment were evident along with the rapid growth of the output value, development of related industries, services and job opportunities. For example, in 1995 the tax levied from purchase and use of cars in Japan alone amounted to 8,254 billion Japanese yen, which accounted for 9.3 per cent of the country's total. Meanwhile, in 1997 the taxes levied from production, purchase and use cars in Germany amounted to 200 billion marks, which accounted for 23.4 per cent of the country's total (International Monetary Foundation, 1996).

Additionally, the increase of car imports and exports through car-trading via the Car City will obviously help the build-up of foreign exchange earnings and in consequence to accommodate the balance of foreign trade of the country as a whole. It can be anticipated that along with further expansion of the world car-market, the total amount of car exports in the world will increase correspondingly. Since the 1990s, the total amount of cars exported around the world has maintained 18.5 million, which accounted for around 40 per cent of the world's total in hand, and the total export value of cars and parts amounted to 500 billion US dollars, which accounted for 10 per cent of the world's total export value, making it one of the most important sectors in exports. In 1998, Germany gained a positive balance in car trading at 78.7 billion marks, which accounted for 60 per cent of the country's total and, again, in 1997 Japan gained a positive balance in car trading at 70 billion US dollars, which accounted for 83 per cent of the country's total, with France at 64.3 billion francs, which accounted for 54 per cent of the country's total. Notably, positive balance in car trading gained by Korea and Spain has played an important role in reducing their deficit in foreign trade. Taking Korea for example, in 1996 Korea's total deficit in foreign trade was about 21 billion US dollars, while the positive balance in car trading gained that the same year was 9.4 billion US dollars (International Monetary Foundation, 1996).

Expansion of Employment The 'third form' of the car-making producer services industry is designed to provide customers with better after-sales service and related activities besides sales, including cultural generalization or popularization of cars, etc. It is clear that the car-making industry cannot be distributed or duplicated in one place randomly; nevertheless, networked car-related service businesses can be distributed and thus create more job opportunities. Nowadays, in developed countries the total output value of the service industry has generated three quarters of the total GDP, and has provided 80 per cent of the total job opportunities in these countries. In the 1990s, the amounts of direct and indirect employment of the car-making industries in the United States, Japan, Germany, France and Korea accounted for 1.6 per cent, 2.8 per cent, 5.5 per cent, 2.8 per cent and 1.5 per cent respectively of the country's total. If adding to this figure job opportunities created by the sale and use of cars, the total amount of employments in the car-making industry and car-related service businesses would increase broadly. Statistically, in accordance with the ratio between direct and indirect employment in the car industry including sales, use of cars and other services, in the United States it was 1:8 (1996), with that of Germany at 1:7 (1997) and Japan at 1:7 (1994). Thus, practically, with the 'third form' of the car-making producer services industry placing more emphasis on car-related services and trade, the growth rate of employment in the service industry will be ten times more than that of the manufacturing industry (Shao, 2002).

Increase of the Output Value of Related Industries   The development of the 'third form' of the car-making producer services industry can also directly promote the growth of related industries through the extension of its industrial chains, or more precisely, rise of an entirely new economy of the car industry. Data from developed countries indicate that there is about 70 per cent of the total output value in the car-making industry actually transferred from other sectors through consumption of the materials and their products. When the output value produced by the car-making industry increases by 1 yuan, it will bring in 0.65 yuan to that of the iron and steel, non-ferrous metals, petrochemical, rubber, kerosene oil, glass, paint, plastic materials, electronic and machinery industries respectively, and 2.63 yuan to that of the tertiary industries of the transportation, infrastructure and urban construction, commerce, tourism, entertainments, banking and financial and other service industries correspondingly (see Table 3.1).

From the above discussion, obviously, the positive effects exerted by the development of the 'third form' of the car-making producer services industry on the overall growth of the service industry are evident. In this, through the extension of the industrial chains of the car-making industry, or again, rise of an entirely new economy of the car industry, it will be beneficial for many related industries in this regard. By statistics, in developed countries, normally, it will pay for around 40 per cent of the total expenditure relatively to service industries when buying a car, including shares of the financial institution, insurance company, legal consultation, industrial advice service, R&D, design, advertisement and so on. Statistics also show that for total car business the ratio of the profits from sales of cars, parts and maintenance is about 2:1:4 (Guo, 2001). Consequently, it can be concluded that it is

Table 3.1   **Demands to related industries from the net increase of the output value of the car-making industry**

Country	Net increase of the output value of the car-making industry	Demands to related industries
The United States (1994, 100 million USD)	874	2250
Japan (1994, trillion Yen)	10	27.6
Germany (1995, 100 million DM)	934	1,800
France (1994, 100 million Franc)	985	2,287

Source: Zhang, Y. X. and others 2001 Effects of development of the car making industry on the national economy, in Report on China's Industrial Development, China Light Industry Publishing House: p.159

also clear that the most part of the profits in car business does not come from the sale of cars, but from after-sales and many other derivative services.

Possible Side Effects However, it should be noticed that there might also exist side effects in this regard which cannot be neglected carelessly, e.g. high sales in the car-selling market on one hand and waste of the production capability on the other. Explicitly, these may include long period of returns and inscrutable risks in financial balance and waste of the production capacity and relevant resources.

Long Period of Returns and Inscrutable Risks in Financial Balance By now, there has been in total 50 billion yuan invested in the Car City in Shanghai, which is predicted to be the biggest car city in Asia. Moreover, recently, the Beijing Municipal Government has planned to spend more than 10 billion yuan in building up the Shunyi Car City in Beijing over eight years. Meanwhile, officials of the First Auto Plant in Changchun of China declare that they plan to build up a world-class car city in Changchun whilst the government of Guangzhou in southeast China plans to create a 'Chinese Detroit'. The impulse for the building up and rapid expansion of car cities around China certainly indicates a lack of overall planning, and will possibly result in the serious problem of a bubble economy. In this, there has been evidence of overspreading of car-making projects blindly across China, and experts from China have pointed out that this may result in an imbalance between supply and actual demand (Jia, 2003).

In Shanghai, meanwhile, except for the Shanghai International Car City in Anting, there is a 'Car-trading Avenue' in Zhabei in the north of the city of Shanghai and it wants to attract leading car-sellers from both the domestic and overseas companies into this area. Alongside this avenue, there are many spare parts shops, car hotels, restaurants, clubs, insurance offices and cultural centers in existence (Zhang and Wu, 2003, p. 17; and similarly hereinafter).

Also, another large-scale car city called the 'Shanghai Car Market', with functions of car selling and services, is under construction in the southeast of Shanghai. By plan, with a total investment of 2.8 billion yuan, the Shanghai Car Market will cover a land space of 600,000 square metres. The investment in the first phase will be 0.4 billion yuan, including an international car exhibition centre, a centre for '4S' brand presentation, a highway exhibition hall, a multiple-purpose exhibition hall, a car information centre, a car logistics centre, a hotel, an international car conference centre, a road test ground, a super class car-trading market and other functional areas.

Besides those mentioned above, there are also some other medium-sized car cities and avenues under construction in Shanghai. These include the Car Avenue of the Wuzhong Road in the west and the Car Culture and Recreation Avenue of the Dongchang Road in the east. Though each of them has its own identity with the Car Avenue of Wuzhong Road putting emphasis on functional car consumption and the Car Culture and Leisure Avenue of Dongchang Road emphasising car-related cultural atmosphere, it is difficult to estimate their effectiveness yet due to the immaturity of the current customers in China.

Although the car industry, in particular the car-making producer services industry, in China is just at the beginning stage compared with that in developed countries, and it is making a good attempt to develop further, it is too anxious and will possibly result in more investment with more risk as the market law indicated previously.

According to the current state of the car market in Shanghai, the chance of a sudden growth in car sales in such a short time is small if there is no great external stimulation. Therefore, it is not yet clear, when investing tens of billions of Chinese yuan to build such cities and avenues, whether the car market will be flourishing or not and how long it will take to see returns on the investment. Add to this the fact that to build a car city normally takes a long period of time, returns will be even longer, it will then possibly result in inscrutable risks in financial balance and thereby a collapse of the market.

Waste of the Production Capacity and Resources At present, although the price of cars has been falling and buying a car costs less than 100,000 yuan in China, to many people a car is still not a 'must have' item. If using the average rate of the annual *per capita* car-related expenditure of the total in the UK of 3.9 per cent (see Table 3.2) to calculate that in Shanghai, a conclusion can be drawn that it is still difficult for an ordinary person to purchase a car at the moment. Therefore, the demand for cars in Shanghai's market is not that optimistic, as many people have estimated previously (see Table 3.3).

Until now, the total annual capacity of the Chinese car industry has been 1.5 million units excluding some small-sized factories, and the actual output for last year was 700,000. In 2004, the total output is expected to be 800,000. Consequently, there has been a large part of the existing capacity set aside. On the other hand, in 2004, the growth rate of China's total car market had reached 30 per cent and, by estimation, in 2005 the total demand for cars in China's market will only reach 1.5–1.6 million. Data show that the 80,000 cars produced in 2003 were actually in excess of demand. This has inevitably led, and will still lead, to waste of the production capacity, resources and inefficiency.

Table 3.2   Annual per capita car-related expenditure of the total in the UK

Year	1988	1989	1990	1991	1992	1993	1994	1995
Percentage (%)	4	3.8	3.8	3.8	3.7	4	3.8	4

Source: UK Statistics Summary, 1991–1997

Table 3.3   Possible annual per capita car-related expenditure in Shanghai

Year	Per capita DPI * in Shanghai (yuan)	Possible per capita car-related expenditure (yuan)
2001	12,883	502
2002	13,250	517

Source: calculated from data in table 1
*Note: DPI refers to disposal income

Locational Factors of the 'Third Form' of the Car-making Producer Services Industry in Shanghai

At present, there have been 10 large car cities settled around the world in total. Among them, that in Detroit in the United States is the top, with those in Toyota City in Japan, Stuttgart in Germany, Torino in Italy, Wolfsburg in Germany, Tokyo in Japan, Paris in France, Birmingham in Britain, Russelsheim in Germany and 'Biyanggu' in France following in sequence. The construction of the car cities and their development marks rise of the new generation of the car-making producer services industry in the contemporary world.

Compared with the ones in western countries, with respect to locational factors of the 'third form' of the car-making producer services industry in Shanghai, there are some similarities and differences.

Similarities to the Western Examples

Large-scale Production Capability of Cars
Similar to the western examples, Shanghai maintains a large-scale economy of the car-making industry and, undoubtedly, it is the most important factor in terms of development of the 'third form' of the car-making producer services industry in China. For many years, the total output of cars in Shanghai has maintained the first place in china's overall market, approximately twice as much as that in Jilin which ranks second in China. Thus, Shanghai is one of the places in China with large-scale production capability for cars (see Table 3.4).

Particularly, as one of the top three auto groups in China, the Shanghai Automotive Industry Corporation Group (SAIC) is located within the Shanghai International Car City. It is engaged largely in manufacture, R&D, trade, finance and other services of cars, trucks, tractors, motorcycles and spare parts. In 2003, a total of 97.2 billion yuan of sales revenue was gained by the SAIC, with total industrial output value at 157.8 billion yuan, sales of vehicles at 800,000 units in which cars at 590,000 (Ding, 2003). By its target, the SAIC will produce 740,000 cars in 2004 so as to maintain its first position in China's total market. This is one of the key reasons why the third form of the car-making producer services industry can develop in Shanghai in advance of elsewhere in China.

Table 3.4   Car output of the top five provinces and municipalities in China

	2001		2002		2003	
	10 thousand	proportion (%)	10 thousand	proportion (%)	10 thousand	proportion (%)
Shanghai	28.88	41.0	39.05	36.8	58.83	28.4
Jilin	15.54	22.1	22.69	21.4	35.07	16.9
Guangdong	5.57	7.9	5.90	5.6	18.33	8.9
Tianjin	5.10	7.2	9.21	8.7	17.25	8.3
Hubei	7.22	10.3	9.74	9.2	13.23	6.4
Total	70.36	100	106.24	100	206.89	100

Source: Wang, Z.H. 2003 Rising of the Shanghai car-making industry, Website of Shanghai Statistics Bureau.

External Economies   Shanghai is situated at the meeting point of the Yangtze Economic Zone and the Coastal Zone in China. Recently, moreover, this locational advantage has been enlarged enormously thanks to the further opening up and great performance of the economy of the Yangtze Valley, particularly the Yangtze Delta Area. The Shanghai International Car City is located in the west of Shanghai and is one of the town areas that Shanghai plans to give priority to develop. Close to the Wusong River in the south, it is 32 kilometres away from the central city of Shanghai and 25 kilometres from the Shanghai Hongqiao Airport, the Shanghai Railway Station and the Zhanghuabin International Containership Terminal. There are, in total, more than 20 roads and highways connected to the central city of Shanghai and its surrounding areas. Additionally, the Shanghai-Beijing and Shanghai-Hangzhou railway lines pass through this area.

The advantages of its geographical location and a convenient transportation network bring to the Shanghai International Car City an extensive hinterland and good external economies. In this, the Yangtze Delta Area holds 5.89 per cent of China's population, on 1.04 per cent of its land. In 2002, it gained 18.25 per cent of the country's total GDP (Gross Domestic Products), 19.26 per cent of the country's total revenue and 24.98 per cent of the country's total gross industrial output value. That same year, the realized FDI reached 17.85 billion US dollars, which accounted for 32.45 per cent of the country's total. A more encouraging phenomenon is that, for nearly 20 years, it has been maintaining a two-digit increase in terms of GDP. In 2003, the Yangtze Delta Area's total GDP was 2,277.5 billion yuan, which increased to 365 billion yuan over the previous year and accounted for 19.5 per cent of the country's total (Shen, 2004). The rapid growth of the external economies brings about huge demand for cars and related services from the Shanghai car market, including needs of transportation, trade and financial services, etc. and may promote a boom of the 'third form' of the car-making producer services industry in Shanghai in the future.

Layers of Investment and the Level of Relative Industries Layers of investment refer to the successive cycles of economic development in a particular place or a region (Massey, 1995). After nearly 20 years of effort, and because of the effect of layers of investment, a car-making industrial system, with combinations of the large, medium and small-sized enterprises and the core and correlated industries, has been eventually formed in Shanghai. In particular the correlated industries have been relatively mature. This has laid a solid foundation for the rise and development of the 'third form' of the car-making producer services industry in Shanghai. Because of this, the related service industry has recently made great progress in Shanghai, with a highly concentrated network of car-trading and service businesses. Based on this, therefore, it is easy for Shanghai to use the existing advantages and establishments to develop its new and rising model of the car-making producer services industry.

Capabilities of R&D Capabilities of R&D in car industries and relevant businesses are among the main foundations too in terms of development of the 'third form' of the car-making producer services industry in the world. Specifically, by now there have been more than 1,600 R&D centres of different kinds in Shanghai, with a total of 1.4 million professionals and 39 universities involved.[4] By plan, in the Shanghai International Car City, there will be a Tongji Auto Institute to set up. Within this institute, there will be a National Car-engineering Centre and a world-class centre for modern businesses in car-trading and management. It will then provide for professionals in this regard for development of the 'third form' of the car-making producers service industries in either Shanghai or around the country.

Differences from the Western Examples

Production Cost As far as the car-making industry and related services are concerned, compared with western countries, the average level of education of the laborers in Shanghai is higher while the average cost is lower. Laborers with a relatively high level of education and lower wages help to conduct manufacturing and service activities that need technology and large-scale economies. For this reason, Shanghai has more advantages in this respect than places in developed countries, and possesses favorable conditions to be one of the international bases of the car-making producer services industry with low cost.

Market Potential It should be mentioned that although the car-making industry in Shanghai is highly competitive in China, at present, local consumption of cars has been relatively lower than in some other cities. In contrast, by now there are in total more than 1 million privately-owned cars respectively in Beijing and Chengdu and

---

4   Same as note 2.

more than 500,000 in Guangzhou while there are only 200,000 in Shanghai. In 2003, the *per capita* GDP of Shanghai was 4,600 US dollars, while 5,642 US dollars when calculated on the basis of the regular population, which approximately matched the level of Japan in the 1960s where the car-making industry in Japan gained high-speed development. This, to a certain extent, indicates the tremendous potential of the actual car market of Shanghai.

Furthermore, it can be expected that an even bigger market will be formed in areas around Shanghai in the forthcoming years beside the Shanghai local market. With further development of its economy, Shanghai will exert more effects on adjacent areas. Meanwhile, quick stepped urbanization processes in the surrounding areas have brought in and will still bring in expansion of the external economies and more opportunities in the development of the 'third form' of the car-making producer services industry in Shanghai.

Conclusions

From the above discussion, it can be concluded that: i) the 'third form' of the industry recognizes the interconnections among marketing, design, leasing and financial services and promotes the extension of the industrial chains of the car-making industry, or more precisely, the rise of an entirely new economy of the car industry as a whole; ii) compared with the traditional car industry, besides car-selling and making, it characterizes itself with its own identities of more and comprehensive functions on maintenance, car-related cultural generalization, R&D activities and many other relative services in one place; iii) through emphasizing technology innovation, it provides the contemporary world with more intensive and thoughtful facilities and comprehensive and humanitarian services, and helps to meet public demands to the maximum; iv) therefore, it may strengthen the vitality and competitiveness of the regional economy and promote further development of the New Economy and New Economic Spaces around the world; v) however, excessive and overwrought investments in this field might be harmful to sustainable development of the regional economy and healthy development of the industry itself; vi) locational determinants of the 'third form' of the industry are mainly concerned with the large-scale production capability of cars, external economies, layers of investment and the level of relative industries, capabilities of R&D, production cost and market potential and, it tends to concentrate related industries in urban regions with favorable conditions of the above factors; vii) the case of Shanghai suggests that this is one of the examples of path dependence in regional economic development and, through its growth, it can be estimated that the overall structure of the car industry in the region will be improved to a high level of certain extent; and viii) the development of the 'third form' of the industry in Shanghai, or more broadly in China, will inevitably exert impacts on the world economy in general and the New Economy in particular. Prospectively, and hopefully, it will then become one of the foundations of the new economy that needs more attention, as geographers seek to understand the dynamics of 'new economic spaces' and order in the contemporary world.

It should be mentioned, however, that although many efforts have been made in this chapter, to a certain extent, because of its nature the complexity and actual role of this industry in creating innovative networks in both the regional and global terms are still not clear and this may lead to a future research direction in this regard.

References

Arguelles, Santiago R. Martinez and Morollon, Fernando Rubiera (2004) 'Outsourcing patterns of advanced business services in the Spanish economy: explanatory elements and estimation of the regional effect', WPSSS13, School of Geography, Earth and Environmental Sciences, The University of Birmingham.

Bryson, J.R., Daniels, P.W. and Warf, B. (2004), *Service Worlds: People, Organizations, Technologies*, London: Routledge.

Cao, P. and Zhao, W. (2003), 'Future-oriented car industry of China', *Chinese Quality*, 27 March, 2003.

Daniels, P.W. (ed.) (1991), *Services and Metropolitan Development: International Perspectives*, London: Routledge.

Daniels, P.W. and Moulaert, F. (eds) (1991), *The Changing Geography of Advanced Producer Services*, London: Belhaven.

Daniels, P.W. (1993), *Service Industries in the World Economy*, in IBG Studies in Geography, Oxford: Blackwell.

Daniels, P.W., Hutton, T.A. and Ho, K.C. (eds) (2005) *Service Industries, Cities, and Development Trajectories in the Asia Pacific*, London: Routledge.

Daniels, P.W., Illeris, S., Bonamy, J. and Philippe, J. (1993), *The Geography of Services*, London: Cass.

Ding, B. (2003), 'Another rich harvest year of the Shanghai Automotive Industry Corporation Group', *Jiefang Daily*, 30 November, 2003.

Gibbons, M. et al. (1994), *The New Production of knowledge*, London: Sage.

Guo, K.S. (2001), *Gap and Catching up: Comparisons between Chinese Industry and that in the Advanced World*, Beijing: City Publishing House: p 290.

Industrial Development Bureau of the United Nations (1997), *Statistical Yearbook of the World Industries*, New York, UN.

International Monetary Fund (1996), *Statistical Yearbook of State Revenues*, Washington DC; IMF.

Jia, X.G. (2003), 'Hot and in void in car-related investments', *Economic Consultation*, 9 September, 2003.

Massey, D. (1995), *Spatial Divisions of Labor*, 2nd ed., London: Macmillan.

Marshall, J.N. and Wood, P.A. (1995), *Services and Space: Key Aspects of Urban and Regional Development*, Harlow: Longman.

Nigel Thrift (2002), 'Producer services', in R.J. Johnston, D. Gregory, G. Pratt and M. Watts (eds), *The Dictionary of Human Geography*, Fourth Edition, Oxford: Blackwell.

Shao, Q.H. (2002), 'The car industry will be a strong propeller of the Chinese tertiary industry', *Car China*, 28 October, 2002.

Shen, Y.F. (2004), 'The Current state, problems and direction of the process of regional integration in the Yangtze Delta Area', *China Business Monthly* (2).

Wu, Y. (2003), 'No. 3 of the series reports: sustainable development of the car industry in China', *Car Business Week*, Chinese Car Industry, 31 October, 2003.

Zhu, Z.Q. (2003), 'Ten keywords of the Chinese car market in 2003', *Southern City*, 23 December, 2003.

Zhang, X.Y. (2003), 'An analysis of the progress of the Chinese car industry', *Car Sina*, 25 November, 2003.

Zhang, Z.B. (2002), *Comparative Advantages and the China's Car Industry: Policy, Mode and Strategies*, Research Centre for Chinese Economy, Beijing University.

Zhang, Y.L. and Wu, L.Y. (2003), 'A new model of the car-selling business: the car market makes rise of the 'third form' of the car-making producer services industry in Shanghai', *Car Business Week*, Jiefang Daily, 22 October, 2003.

Chapter 4

Metropolitan Cities as the Innovation Centers of Knowledge-Intensive Business Services: The Case of Seoul in Korea

Ji-Sun Choi[1*]

Introduction

Industrial restructuring and innovation have been at the center of discussion since the late 1990s, especially in Korea after the IMF turmoil, because they were believed as absolutely important keys to national competitiveness. However, the role of service industries in the innovation process and the innovation of service industries has continued to be neglected, despite the high proportion of service industries in gross domestic production. Top priority has been put on the innovation capacity of manufacturing industries. In Korea, where the population is about 49 million people, almost 10 million people worked for establishments related to service industries at the end of 2002. The share of employees in service industries is 67.8 per cent, while that in manufacturing industries is only 23.2 per cent. However, the focus of the research on technological innovation has been on manufacturing industries in Korea (i.e. Kim, S.K., 1999; Kim, S.K, 2004; Chung et al., 2000).

As for the importance of service innovation, this chapter focuses on the two key aspects: first, innovation by knowledge-intensive business services (KIBS) and, second, its geographical clustering in metropolitan cities, especially in Seoul, which is the largest metropolitan city in Korea. It is known that more firms in the Seoul Metropolitan Area (SMA) tended to succeed in the technological innovation of service sectors than those in the regions outside the SMA (Choi, 2005a).[1]

This chapter begins with a literature review concerning service innovation and its location. Secondly, it describes the geographical distribution of KIBS in Seoul as well as in Korea generally. Thirdly, it offers a statistical analysis of the characteristics of the technological innovation by KIBS in Seoul by comparing it with non-Seoul

---

1   SMA is the abbreviation of the Seoul Metropolitan Area, which includes Seoul Metropolitan City, Incheon Metropolitan City, and Gyeonggi-do (province).

areas (the provinces).[2] Finally, some policy implications for ways to develop KIBS in Seoul and the provinces are considered.

Service Innovation and Location

The literature on service innovation has been gradually increasing across the globe (for example, Hauknes, 1998, Sirilli and Evangelista, 1998, Hertog, 2000, Metcalfe and Miles, 2000, OECD, 2001, Nählinder 2002, Nählinder and Hommen, 2002, Hollenstein, 2003, Lee et al., 2003, Um and Choi, 2004a, 2004b). Services are not necessarily passive elements in the innovation system or simple consumers of innovation that are locked in to a dominant manufacturing innovation system (Howells, 2001: 67). Services perform important roles in the innovation process and actively create new value added during the innovation process. A positive relationship between accessibility to business service and industrial innovation has been empirically demonstrated (MacPherson, 1997). In particular, the growth of KIBS is often considered as a key success factor to a client (often manufacturing) firm's innovation activities. More recently, KIBS are regarded as co-producers of innovation, based on a symbiotic relationship with client firms (Hertog, 2000). The view is gradually gaining ground that the self-innovative capabilities of KIBS are just as important as the innovation-facilitating capabilities for client firms as co-producers. According to Muller and Zenker (2001) KIBS do not only influence the innovation of client firms, but also benefit from the interactions with their clients, thus constituting a virtuous circle. KIBS help to strengthen the innovative capacities of client firms and gain stimuli for their own innovations through knowledge transmission and re-engineering processes.

Recognition of the importance of KIBS innovation often leads to a controversy about the importance of its technological innovation. It has always been believed that manufacturing firms were more oriented towards technological innovation than service (or KIBS) firms. The latter were usually connected to non-technological innovation.[3] However, successful manufacturing firms started to recognize the importance of non-technological innovation, and successful service firms also recognized the importance of technological innovation as a way of satisfying

---

2   The provinces, in this study, are defined as the non-Seoul areas, which encompass all the regions outside the Seoul Metropolitan City.

3   A four-dimensional model of service innovation is suggested by Hertog (2000). The four dimensions include the introduction of new service concept (conceptual innovation), new client interface (client-interface innovations), new service delivery system (delivery system and organization innovations) and technological options (technological innovations). Although technological dimension is not always used for service innovation, technology is connected to service innovation in many cases in practice. The influence of technology in service innovation varies ranging from a facilitating factor to technology-driven innovation. The rapid adoption of new ICTs in service sectors enhances the technology-intensity of service firms in the era of knowledge-based economy.

consumers' needs (Howells, 2001: 65). A sharp increase in the R&D expenditure of total business R&D in service industries is clear evidence for the growth in the importance of the technological innovation in service industries. The increasing share of service R&D results from various factors: these include better measurement of R&D in services, more research by service firms, and increased outsourcing (Edwards and Croker, 2001: 9).

From a geographical perspective, KIBS tend to be more concentrated in large cities than other types of service firms as well as manufacturing firms. The location of business services is differentiated by the complexity of business services related to the product life cycle of client firms (Cappellin, 1989). 'Orientation services' (such as R&D services or advanced consulting services etc.) are needed for the client firms in the initial phase of the product life cycle and 'planning services' (technical, organizational, financial, and marketing services etc.) are required for those in the development phase. 'Programming services' (routine services about daily organization etc.) fulfil the needs of client firms in the maturity phase of the product life cycle. Orientation service firms tend to localize in the large metropolitan areas and planning service firms are likely to be located in medium sized urban centers. Programming service firms can be localized in even smaller urban centers of peripheral regions close to production plants.

The concentration of KIBS in large metropolitan areas is often explained with the ease of innovation. Muller and Zenker (2001) argue that proximity (geographical, social, cultural, etc.) is of great use to exchange tacit knowledge between KIBS and their clients, regardless of the unprecedented development of ICT.[4] This is in line with Feldman (1994) who explains the reasons for the importance of geographical clustering for innovation. The timely exchange of information and the accumulation of tacit and codified knowledge is an essential part of innovation. Various information sources such as related firms, university, specialized business services, and client firms are the key to innovative success because they reduce the uncertainty of innovation. This is one of the reasons that innovation is expected to be associated with geographical clustering (Feldman, 1994: 27).

It may be inferred then that the more important innovation, specifically technological innovation, becomes for KIBS, the higher the geographical tendency toward clustering in large cities. The geographical clustering of innovation-friendly KIBS in large cities encourages them to sell their products or services to remote customers with the help of the development of ICTs (Illeris, 1989). It may also be

---

4   The exchange of tacit knowledge is sometimes considered more important in the relationship with different service firms than with client firms (Cappellin, 1989). The purchase of inputs by service firms needs more frequent face-to-face contact and more transaction costs because the sources of inputs are diversified. Therefore, even then services can be provided to the client firms in remote areas, the cost required in purchasing inputs to service firms results in the concentration of service firms in large metropolitan areas (Cappellin, 1989: 268). In addition, the importance of face-to-face contact is emphasized in the case of advanced services, which may contribute to the concentration of high-order services in the opposition to the diffusion of low-order routinized and standardized services.

closely linked with the division of labor between the centralization of important business services and the decentralization of routinized and back-office services into peripheral regions.[5] The role of innovation-friendly KIBS as basic activities is likely to be strengthened as centralization is enhanced, together with the unprecedented development of ICTs.

Data and Definitions

This chapter draws on data from the *Korean Innovation Survey 2003: Service Sector*, which was performed by the Science and Technology Policy Institute (STEPI) and the Korea Information Strategy Development Institute (KISDI) in 2003, following the guidelines of the OSLO manual (OECD, 1997).[6] The number of respondent service firms is 2,000, which is about 10.2 per cent out of the total number of establishments (19,603), as recorded by the *2002 Census on Basic Characteristics of Establishments* by the Korea National Statistical Office (KNSO, 2003).

Three kinds of two-digit level KIBS industries are regarded as narrowly defined KIBS for the purposes of this study. They are Computer and Related Activities (Computer services, KSIC=72), Research and Development (R&D services, KSIC=73), and Professional, Scientific and Technical Services (P-KIBS, KSIC=74).[7] They all belong to the category of Business Activities (BA, KSIC= M) in the 1-digit level of KSIC.[8] Although there still remain many industries belonging to KIBS, these three industries are often considered as the most representative KIBS industries. They are sometimes sub-divided into two types, i.e. technology-based KIBS (T-KIBS) and professional services (P-KIBS). T-KIBS includes computer and R&D services and is known to be more dependent on new technology and more active in innovation than P-KIBS (Nählinder, 2002). The number of the narrowly defined KIBS firms out of 2,000 total respondents is 743, which is 14.6 per cent out of the

---

5   Coffey and Polèse (1989) put together several research that shows some empirical evidences that the employment growth in business services is in the form of expanded suburbanization (i.e. deconcentration), rather than a true decentralization into lagging regions. According to them, the concentration of business services in a few large cities is explained with some factors including urbanization economies, a marked concentration of corporate headquarter and control functions, intrafirm functional separation between front and back office function of business services with the help of the development ICT.

6   Please refer to Um and Choi (2004a; 2004b) to find out the detailed analysis of the survey results.

7   Legal, Accounting, and Tax Preparation Services (KSIC=741) are excluded from P-KIBS category in the 2003 Korean Innovation Survey.

8   In fact, there is the other two-digit level industry belonging to Business Activities category. It is called 'other business service activities (KSIC=75)'. The industry is not dealt with importantly in this chapter because they are not much related to innovation processes. It is composed of miscellaneous business support services, which include facilities support and employment Services, investigation and security services, building-cleaning services and so on.

total number of KIBS firms. In terms of the three sub-industries, the sample share in R&D services is higher than the other two industries, because of the very small total number of R&D service firms.

The working definitions in the survey are set up following the guidelines of the OSLO manual (OECD, 1997). Technological innovation is based on using radically new technologies or combining existing technologies in new uses or can be derived from new knowledge.[9] Technological innovation consists of technological product (or service) innovation and technological process innovation. Technological product innovation is again divided into two sub-types such as creating technologically new products and technologically improved products. Technological process innovation is related to adopting technologically new or improved production methods.

Geographical Distribution of KIBS in Korea

A significant change in industrial structure caused by voluntary or compulsory relocation of manufacturing firms to suburban areas leads to an extreme concentration of control function and advanced business services such as corporate headquarters, engineering services, software services, design and engineering services in Seoul (Park and Nahm, 1998). The concentration of KIBS in Seoul has accelerated since 1995 in parallel with the advent of a knowledge-based economy. The development of knowledge-intensive services in Korea has resulted in the concentration of those firms in the SMA. According to Kim, Y.S. (2003), using a broad definition the proportion of employees in the SMA in knowledge-intensive service industries amounts to 71.6 per cent in 2000. In particular, the proportion of employees in KIBS establishments (KSIC=72, 73, 74) in Seoul is 55.3 per cent (292,911) and that of the number of establishments is 42.4 per cent (26,619), while the share of employees is 29.3 per cent in all service industries and 16.8 per cent in manufacturing industries in 2002.[10]

Location quotients (LQs) of manufacturing industries, total service industries, KIBS firms, and the incorporated KIBS firms further demonstrate the extreme concentration of KIBS in Seoul.[11] There is a clear contrast in the LQ values between manufacturing industries (LQ=0.65) and KIBS (LQ=2.12) (Figure 4.1).

---

9 For more information on the definitions and examples of the concepts, please refer to OECD (1997).

10 The ratio of the employees in all the establishments belonging to Business Activities (KSIC=M; 72, 73, 74, 75) is 50.9 per cent (435,898) and that of the number of establishments is 39.8 per cent (31,719) in Seoul as of 2002.

11 The concept of incorporated companies differs from that of establishments. The term 'establishment' covers all units situated in a single location (such as individual shops, offices, banks, agents, schools, hospitals, hotels, restaurants, all kinds of academic institutes, churches, Buddhist temples, public organizations, social welfare facilities, etc) operated under a single ownership or Control. Incorporated companies include joint-stock companies, limited

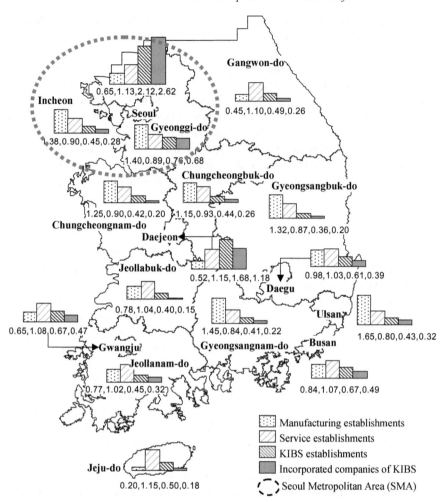

Figure 4.1   LQ's, by region and industry, 2002

One of the noteworthy findings about the development of KIBS in Korea is that most of the metropolitan cities with a population of more than 1 million, except for Seoul, do not reveal any clear concentration of KIBS (Figure 4.1). Busan, Daegu, Incheon, and Ulsan Metropolitan Cities have higher LQ values for manufacturing industries than for KIBS. The analysis by industry shows the main characteristics of the KIBS firms concentrated in Seoul and other metropolitan cities (Table 4.1). Seoul is notable for its major concentration of computer services, although the other industries are also more concentrated in Seoul than in the other metropolitan cities.

companies, joint-stock limited partnerships or unlimited partnerships established under the regulations of commercial law and can be composed of more than one establishment.

Table 4.1.   LQs, by industry: metropolitan cities, 2002

Metropolitan Cites	Population	Computers services	R&D services	P-KIBS	Other services
Seoul	10,280,523	3.11	0.98	1.89	1.67
Busan	3,747,369	0.40	0.29	0.86	1.02
Daegu	2,540,647	0.36	0.27	0.79	0.77
Incheon	2,596,102	0.16	0.35	0.60	0.68
Gwangju	1,401,525	0.39	0.26	0.89	1.19
Daejeon	1,424,844	0.87	7.41	0.93	1.28
Ulsan	1,070,277	0.11	0.08	0.64	1.82

Note: LQs are calculated using employment in total national establishments as the base value.

Source: calculated by the author with the data from KNSO (2003).

Table 4.2   Regional comparison of the rates of technologically innovating firms, by industry, 2001–2002

(Unit: number of establishments, %)

Industry	Region		Total	Chi-square Exact S. 2-sided
	Seoul	The provinces		
Total	225/477	71/266	296/743	0.000
	47.2	26.7	39.8	
Computer services	139/238	24/57	163/295	0.037
	58.4	42.1	55.3	
R&D services	31/49	24/44	55/93	0.408
	63.3	54.5	59.1	
P-KIBS	55/190	23/165	78/355	0.001
	28.9	13.9	22.0	

Source: calculated by the author with the data from Um and Choi (2004a).

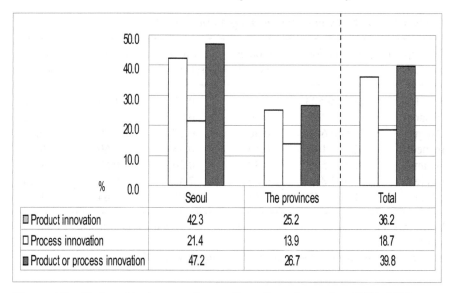

	Seoul	The provinces	Total
▣ Product innovation	42.3	25.2	36.2
▢ Process innovation	21.4	13.9	18.7
▪ Product or process innovation	47.2	26.7	39.8

Figure 4.2   Rates of technology innovationg firms by region, 2001–2002

Daejeon Metropolitan City, with the second highest LQ value for KIBS, is extremely specialized in the agglomeration of R&D services. This is because the central government has strategically developed the city as a center for R&D facilities and it has been the most advanced R&D cluster in Korea since the 1970s (Choi, 2005b). The remaining metropolitan cities mainly rely on the development of miscellaneous business support services, which include facilities support and employment services, investigation and security services, building-cleaning services and so on. Most of these are usually unimportant in relation to innovation processes.

Characteristics of Technological Innovation in Seoul

Rates of Technologically Innovating Firms

The survey results demonstrate a significant difference in the rates of technologically innovating firms between Seoul and the provinces (or the non-Seoul area) at the 0.05 statistical significance level (Figure. 4.2 and Table 4.2). In terms of technological innovation, the share of product or process innovators during 2001–2002 is 47.2 per cent in Seoul but only 26.7 per cent in the provinces. However, even though the proportion of technologically innovating firms in Seoul is much higher than in the provinces, it is not as high as in the EU countries. According to the Third Community Innovation Survey (CIS III) of innovation during 1998-2000, the average technological innovation rate by KIBS with almost the same industrial coverage, including NACE divisions 72 (computer activities), 73 (R&D), 74.2 (engineering activities and consultancy), 74.3 (technical testing and analysis), among EU countries is approximately 57 per cent (EU, 2004).

The concentration of technologically innovating firms in Seoul is also clearly revealed by an analysis based on the three sub-industries (Table 4.2). However, the degree of the concentration is differentiated by the sub-industries. Whereas the rates of technologically innovating firms in computer services and P-KIBS firms are significantly higher in Seoul than in the provinces at the 0.05 statistical significance level, the outcome for R&D services is not much different by region. This seems related to the fact that Daejeon Metropolitan City is the most specialized area for advanced R&D facilities.

Main Sources of Technological Innovation

According to the survey result, KIBS firms in Seoul tend to rely mainly on intramural sources rather than extramural ones in the process of technological innovation (Table 4.3). The average share of new (or improved) products, which were created mainly

Table 4.3   **Average ratio of internal development of technologically new (improved) products, by industry, 2001–2002**

Industry	Internal	Cooperative	Outsourcing	Others
Computer	83.2	10.9	5.2	0.7
R&D	86.4	7.7	4.1	1.8
P-KIBS	73.4	14.0	7.6	5.0
Sub-total	81.4	11.2	5.6	1.9

Source: calculated by the author with the data from Um and Choi (2004a).

Table 4.4   **Average ratio of innovation expenditure by technologically innovating firms, by industry, 2002**

Industry	Intramural R&D	Acquisition of R&D	Acquisition of other external knowledge	Acquisition of machinery and equipment	Training/ Marketing Design etc.
Computer	53.6	8.6	6.2	18.2	13.3
R&D	58.4	5.2	2.7	22.9	10.7
P-KIBS	36.7	8.1	9.9	28.1	17.3
Sub-total	49.7	8.1	6.8	21.4	14.1

Source: calculated by the author with the data from Um and Choi (2004a).

Table 4.5   R&D expenditure, by industry, 2002

(Unit: million KRW, %)

Industry	R&D expenditure	Seoul/Total	
		R&D expenditure	Employments
Total	12,975,354 (100.0)	22.0	26.1
Manufacturing sectors	11,110,689 (85.6)	15.1	16.8
Service sectors	1,161,846 (9.0)	70.9	29.3
Business services*	853,500 (6.6)	76.9	50.9
• Computer services (KSIC-72)	639,414 (4.9)	88.1	81.1
• R&D services (73)	53,217 (0.4)	32.6	25.6
• Other services (74, 75)	160,869 (1.2)	47.2	46.4

Note: Business services include computer services, R&D services, P-KIBS, and other business service activities in this table due to the difficulty of data acquisition on the only P-KIBS.
Source: calculated by the author with the raw data from MOST and KISTEP (2003).

by the use of internal sources, amounted to some 81.4 per cent in the case of KIBS firms in Seoul. In relation to the sub-industries, P-KIBS firms are inclined to be less reliant on the independent development of new or improved products than computer and R&D services. Interestingly, the reliance on the internal resources of firms in computer services in Seoul is prominent when it is compared with the situation in the provinces.

The tendency toward the internal development of technological innovation by KIBS firms in Seoul is also reflected in an analysis of their innovation expenditure (Table 4.4). Technologically innovating firms in Seoul spent about 49.7 per cent of total annual innovation expenditure on intramural R&D programs, whereas the expenditure on the acquisition of R&D and other external knowledge was only 14.9 per cent. In the analysis of the sub-industries, P-KIBS firms in Seoul expended more on the acquisition of diverse types of external knowledge than computer and R&D service firms. The relatively high level of acquiring other external knowledge might be interpreted as indirect evidence that P-service firms in Seoul have easier access to more useful external knowledge sources for technological innovation than provincial P-service firms.[12]

---

12 The analysis result of the innovation expenditure might not be completely matched with that of the tendency of technological innovation because it was just measured on the basis of a year of 2002, although the technological innovation was measured on the basis of the two years from 2001 to 2002.

The general tendency toward technological innovation relevant to internal R&D in Seoul is in line with the concentration of R&D expenditure in the city. As shown in Table 4.5, firms' R&D expenditure is highly concentrated in Seoul and in the SMA. Approximately 70.9 per cent of the R&D expenditure by service industries was made in Seoul in 2002. The proportion of service R&D expenditure in the SMA is over 90 per cent. The proportion of the R&D expenditure by KIBS in Seoul amounts to about 76.9 per cent, the equivalent for firms in the SMA is about 92.3 per cent. The concentration of R&D expenditure by computer services in Seoul is prominent.

External Cooperative Networks for Technological Innovation

This analysis of the ways to develop new products and the innovation expenditure in 2002 shows that the technological innovation by KIBS firms in Seoul is closely related to the internal sourcing or the internal innovative capabilities of individual firms, although P-KIBS firms are relatively less dependent on internal sources than other forms. However, this does not necessarily mean that they have nothing to do with the need for external cooperation in the process of technological innovation. Rather, some previous surveys indicate that the local innovative environment in Seoul is likely to contribute considerably to the technological innovation by KIBS by encouraging external innovative cooperation. For example, Park and Choi (2005) have argued that the development of IT service industries in Seoul, especially in the specific local area of Gangnam, is attributable to the local supportive innovative

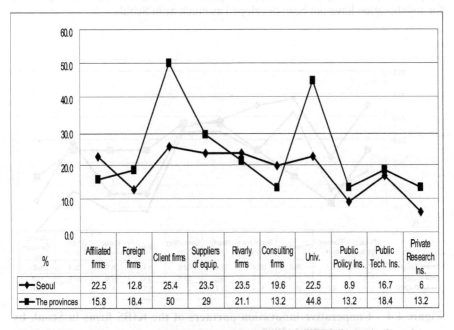

%	Affiliated firms	Foreign firms	Client firms	Suppliers of equip.	Rivarly firms	Consulting firms	Univ.	Public Policy Ins.	Public Tech. Ins.	Private Research Ins.
Seoul	22.5	12.8	25.4	23.5	23.5	19.6	22.5	8.9	16.7	6
The provinces	15.8	18.4	50	29	21.1	13.2	44.8	13.2	18.4	13.2

Figure 4.3   Cooperative innovation partners of the KIBS firms in Seoul

environment that encourages local innovation networks and collective learning processes.

As expected, a considerable number of KIBS firms in Seoul had experience of cooperation with external partners for the purpose of technological innovation. About 45.5 per cent (102 out of 224 firms) of the technologically innovating firms had experienced cooperative activities with external partners in Seoul. The proportion is partly differentiated by type of industry. R&D service firms (54.8 per cent, 17/31) had more experience of collaboration with external partners than computer service firms (44.9 per cent, 62/138) and P-KIBS firms (41.8 per cent, 23/55). In comparison with the CIS III survey results, the average proportion of firms engaged in innovation activity involving collaboration with external partners is about 34 per cent for KIBS in the EU countries (EU, 2004).

However, more importantly, in spite of the high rate of firms with experience of external cooperation, the degree to which cooperative partners contributed to technological innovation was not evaluated so high, at least in this survey. In terms of KIBS firms in Seoul, only about 10–25 per cent of total respondent firms agreed that each cooperative partner contributed to their technological innovation (Figure 4.3). The firms in Seoul do not attach high priority to specific type of cooperation partners. They generally place almost the same amount of importance on cooperative partners whether they are client firms, affiliated firms, suppliers, competitors, consulting firms, or universities. By contrast, provincial firms put special emphasis on client firms or customers and universities. The particular reliance on customers and universities as cooperative partners by provincial firms implies that other potential external partners for collaboration have not been developed as much as in Seoul.

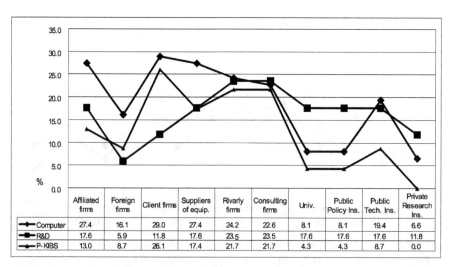

	Affiliated firms	Foreign firms	Client firms	Suppliers of equip.	Rivarly firms	Consulting firms	Univ.	Public Policy Ins.	Public Tech. Ins.	Private Research Ins.
Computer	27.4	16.1	29.0	27.4	24.2	22.6	8.1	8.1	19.4	6.6
R&D	17.6	5.9	11.8	17.6	23.5	23.5	17.6	17.6	17.6	11.8
P-KIBS	13.0	8.7	26.1	17.4	21.7	21.7	4.3	4.3	8.7	0.0

Figure 4.4   Cooperative innovation partners of the KIBS firms in Seoul, by industry, 2001–2002

A detailed analysis of the type of the external partners of KIBS firms by industry leads to an interesting result (Figure 4.4). The firms in computer services and P-KIBS have a similar pattern in terms of the importance of external cooperative partners in the process of technological innovation, although the absolute percentage in the recognition of the importance is normally lower for P-KIBS firms than for computer service firms. The private sector is relatively more conducive to the innovation process of computer and P-KIBS firms than the public sector. On the contrary, in relative terms the R&D service firms place considerable importance on collaboration with the public sector by comparison with computer service and P-KIBS as a whole.

Firms' Internal Attributes Affecting the Technological Innovation of KIBS

Regression analysis has been used to establish the relative importance of different in-firm attributes on the likelihood of engaging technological product or process innovation.[13] A binary logistic regression model is used, because a dependent variable is a binary form that has only two values and the model requires fewer assumptions than original regression models or discriminant analysis. The model is estimated using two steps. First, it is measured on the total number of KIBS respondents and the firms in Seoul and in the provinces, respectively. Then, it is narrowed down to focus on the sub-industries in Seoul and estimated in the model by T-service (representatively, computer services) and P-KIBS.

In the logistic regression model on all the respondents, the number of employees in 2000 (regression coefficient B=0.34), the spatial boundary of the target market (B=-0.44), the year of establishment (B=0.39), and the type of industries (B=1.57) are all statistically related to the occurrence of technological innovation at the 0.05 level (Table 4.6).[14] The KIBS firms that are large employers, cover the whole domestic market, are the most recently established, or related to T-services tend to have a higher level of technological innovation.

For the models by region, the number of employees in 2000 (Seoul B=0.30/the provinces B=0.38) and the type of industries (Seoul B=1.3/the provinces B=1.84) influence the occurrence of technological innovation significantly in the provincial model as well as in the Seoul model at the 0.05 level. Interestingly, the variable

---

13 Setting up the significance level in statistical examination is an important procedure with which a null hypothesis is rejected. It can be differentiated by the purpose of the analysis. The significance level of 0.001 or 0.01 is common in natural science studies, and that of 0.05 is often used in social science studies. The significance level of 0.10 is sometimes applied in social science studies, because many unpredictable phenomena is social sciences are not easily examined through a strict statistical rule.

14 The independent variables with the regression coefficient of more than 0.3 or less than 0.3 are considered important as the significant variables, at least having a moderate relationship with the occurrence of technological innovation, out of the all the variables significant at the 0.05 level.

Table 4.6   Regional comparison of the internal attributes affecting technological innovation, 2001–2002

Covariates (variable name)	Variable type	Seoul		The provinces		Total	
		B	Sig.	B	Sig.	B	Sig.
Employees in 2000 (lna3a)	Ratio (Natural log)	0.3036	0.0027*	0.3785	0.0278*	0.3443	0.0001*
Employees with master degrees in 2002 (a3da3c)	Ratio (%)	0.0072	0.1495	0.0187	0.0071*	0.0108	0.0071*
Target market market (d111a5)	Dummy (l-local, (0=domestic, foreign)	–0.1695	0.5625	-0.3544	0.3316	–0.4376	0.0435*
Product life cycle (a7re2) (1=less than	Dummy (l=less than 5 years)	–0.0647	0.7693	–0.2288	0.5041	–0.0325	0.8584
Firm's age (ageyear2) d1code3	Dummy (l-less than 10 years)	0.41912	0.0735**	0.3983	0.3147	0.3946	0.0465*
Type of industries (dlcode3)	Dummy (l=IT, R&D, 0=professional)	1.3243	0.0000*	1.8404	0.0000*	1.5692	0.0000*
Constant		–2.2448	0.0000*	–3.4267	0.0001*	–2.7320	0.0000*
Included data		441		240		681	
-2 Log Likelihood		560.580		232.450		805.690	
Model Chi-Square (df)		50.501		49.871		117.474	
Sig.		0.0000		0.0000		0.0000	
Model fitness between observation and prediction		65.53%		76.67%		69.60%	

Note: * statistically significant at the significance level of 0.05, ** at the significance level of 0.10.
Source: calculated by the author with the data from Um and Choi (2004a).

Table 4.7 Internal attributes affecting technological innovation in Seoul, by industry, 2001–2002

Covariates	Variable type	Computer services		P-KIBS	
		B	Sig.	B	Sig.
Employees in 2000 (lna3a)	Ratio (Natural log)	.3504	.0158*	.4190	.0175*
Employees with master degree IN 2002 (a3da3c)	Ratio (%)	.0220	.0427*	-.0015	.8570
Target market (d111a5)	Dummy (1=local, 0=domestic, foreign)	−.7137	.0575**	1.0100	.0361*
Product life cycle (1=less than 5 years)	Dummy (1=less than 5 years)	−.7002	.0431*	.2339	.4981
Firm's age (ageyear2)	Dummy (1=less than 10 years)	5972	.0770**	.0769	.8393
Constant		−.7961	.2455	-2.7133	.0013*
Included data		226		173	
−2 Log Likelihood		284.093		203.944	
Model Chi-Square (df)		20.588		9.246	
Sig.		.0010		.0996	
Model fitness between		66.37%		69.94%	

Note: * statistically significant at the significance level of 0.05, ** at the significance level of 0.10.

Source: calculated by the author with the data from Um and Choi (2004a).

on the target market is not statistically significant in two of the regional models, although it is statistically significant in the model using all the respondents.

According to the logistic regression on the firms in Seoul by sub-industry, the number of employees in 2000 (regression coefficient B=0.35), product life cycle (B=-0.70), and the year of establishment (B=0.60) are all related to the occurrence of technological innovation in the computer service industry (Table 4.7). By comparison, only the number of employees in 2000 (B=0.42) and the spatial coverage of the target market (B=1.01) influence the technological innovation by P-KIBS firms in Seoul.

The model means that the larger firms, with more than five years of product life cycle, or which have been established less than ten years ago tend to actively engage in technological innovation in terms of computer service firms in Seoul. On the contrary, the technological innovation of P-KIBS firms in Seoul is much influenced only by employment size or the spatial coverage of the target market. Interestingly, the firms with mainly local markets prove to have a higher probability of performing technological innovation; more than those with spatial coverage extending to domestic and foreign markets. The need for customized services for the local markets is likely to be an important motivation for technological innovation, although the attributes of P-KIBS innovation do need more future research.

Conclusion

This chapter has attempted an investigation of the main attributes of KIBS innovation in Seoul, based on data from the *Korean Innovation Survey 2003: Service Sector* from the perspective of geography and industry. Three types of KSIC 2-digit level industries belonging to Business Activities (KSIC=M) have been selected for this study. This chapter describes the geographical distribution of KIBS firms in Korea and shows the technological innovation rate of KIBS in Seoul by comparing it with the provinces. The latter part statistically examines the main characteristics of the technological innovation of KIBS in Seoul with the focus on the internal and external sources of technological innovation and firms' internal characteristics affecting technological innovation.

There are two issues for future research on KIBS in Seoul based on the main findings of this study. The first is about the policy implication of the tendency toward reliance on internal resources and low importance of the external cooperation of KIBS technological innovation in Seoul. At least, based on this study, technological innovation by KIBS firms in Seoul originates from firms' internal innovative capabilities and external cooperation has an additional importance. Although innovative KIBS firms are considerably clustered in Seoul, they do not seem to enjoy the benefit of clustering with adequate cooperation in terms of essential tacit knowledge exchange for technological innovation. Intensified local networks and improved innovative milieu will create a synergy effect with their in-house R&D capabilities. The evidence on the thickness of innovative milieu, which supposedly encourages external cooperation, is not clearly revealed in this study. In this regard,

Seoul metropolitan government needs to find ways to encourage external cooperation among innovative firms in the region as well as in its neighboring regions in order to improve the quality of knowledge and technology exchange and to facilitate innovation.

Second, the differences between the characteristics of the technological innovation behaviors of P-KIBS and computer and R&D services imply that customized approaches are required to encourage the development of KIBS in Seoul. The P-KIBS firms reveal a lower level of technological innovation than the other types of KIBS even though it is relatively much higher than in the provinces. This study partly suggests that the technologically innovating firms in P-KIBS are less dependent on internal sources than the other types of KIBS. They seem more prepared to engage in cooperative innovation with external partners to compensate for the low intensity of their internal R&D capabilities. Some policy tools are needed to develop the internal as well as external innovative capabilities of the P-KIBS. By comparison, the policies to establish external knowledge transfer by encouraging external cooperation seem more essential for the firms in computer and R&D services if they are to gain a lot of synergy from harmonization between internal and external innovative capabilities and to benefit from the concentration or clustering of KIBS in Seoul.

There is a long way to go with regard to the imperfect methodology used in the analysis of this chapter. Above all, this study just deals with the quantitative dimension of technological innovation by only measuring the rate of the firms that had the experience to succeed in technological innovation and to make external cooperation. However, the quality of technological innovation and external cooperation is more important than their quantitative attributes. The logistic regression models are not perfect partly because multi-collinearity is not completely removed in spite of the effort to achieve this. Furthermore, the study of technological innovation in Seoul is likely to be more valuable when the roles of the city as an innovative milieu and as the main facilitator of innovation are dealt with in greater detail. Hence the importance of adding qualitative measures to the future research agenda.

References

Cappellin, Riccardo (1989), 'The diffusion of producer services in the urban system', *Revue d'Economie Régionale et Urbaine*, **4**, 641–61.

Choi, Ji-Sun (2005a), 'Analysis on the technological innovation and cooperative networks of service industries by region (in Korean)', **40** (1), *Journal of the Korean geographical society*, 63–77.

Choi, Ji-Sun (2005b), 'Potential and limitation of new industrial policy in Korea: fostering innovative clusters', in Le Heron, R. and Harrington, J.W. (eds), *New Economic Spaces: New Economic Geographies*, Hampshire: Ashgate.

Chung, S., Ahn, D. and Lee, J. (2000), *Major Sectoral Innovation System in Korea* (In Korean), Seoul: Science and Technology Policy Institute.

Coffey, William J. and Polèse, Mario (1989), 'Producer services and regional development: a policy-oriented perspective', *Papers of the Regional Science Association*, **67**, 13–27.

Edwards, Mike and Croker, Michelle (2001), 'Major trends and issues', *Innovation and productivity in services*, Paris: OECD.

EU (2004), *Innovation in Europe: Results for the EU, Iceland and Norway*, Luxemburg: European Commission.

Feldman, Maryann P. (1994), *The Geography of Innovation*, Dortrecht: Kluwer Academic Publishers.

Hauknes, Johan (1998), *Innovation in the Service Economy*, Oslo: STEP group.

Hertog, Pim Den (2000), 'Knowledge-intensive business services as co-producers of innovation', *International journal of innovation management*, **4** (4), 491–528.

Hollenstein, H. (2003), 'Innovation modes in the Swiss service sector: a cluster analysis based on firm-level data', *Research policy*, **32**, 845–863.

Howells, Jeremy (2001), 'The nature of innovation in services', *Innovation and productivity in services*, Paris: OECD.

Illeris, Sven (1989), 'Producer services: the key sector for future economic development?', *Entrepreneurship and Regional Development*, **1**, 267–274.

Kim, Seok-Kwan (1999), *The Patterns and Directions of Technological Innovation in Footwear Industry* (In Korean), Seoul: Science and Technology Policy Institute.

Kim, Young-Soo (2003), *Regional Development of Knowledge-intensive Services in Korea and its Policy Implication* (in Korean), Seoul: Korea Institute for Industrial Economics and Trade.

Kim, Seok-Kwan (2004), *Innovation Patterns and Strategies of Pharmaceutical Industry* (In Korean), Seoul: Science and Technology Policy Institute.

KNSO (Korea National Statistical Office) (2003), *2002 Census on Basic Characteristics of Establishment*, Daejeon: KNSO.

Lee, K.R., Shim, S.W., Jeong, B.S. and Hwang, J.T. (2003), *Knowledge Intensive Service Activities in Korea's Innovation System*, Seoul: Science and Technology Policy Institute.

MacPherson, Alan (1997), 'The role of producer service outsourcing in the innovation performance of New York State manufacturing firms', *Annals of the association of American geographers*, **87** (1), 52–71.

Metcalfe, J.S. and Miles, I. (eds) (2000), *Innovation Systems in the Service Economy*, Boston: Kluwer academic publisher.

MOST (Ministry of Science and Technology) and KISTEP (Korea Institute of S&T Evaluation and Planning) (2003), *Report on the Survey of Research and Development in Science and Technology*, Gwacheon: MOST.

Muller, Emmanuel and Zenker, Andrea (2001), 'Business services as actors of knowledge transformation: the role of KIBS in regional and national innovation systems', *Research policy*, **30**, 1501–1516.

Nählinder, Johanna (2002), Innovation in knowledge intensive business services: State of the art and conceptualizations, Presented at the SIRP seminar, Linköping, 15 January 2002.

Nähinder, J. and Hommen, L. (2002), 'Employment and innovation in services:

knowledge intensive business services in Sweden', paper presented at the final meeting and conference of AITEG., London, 18–19 April 2002.

OECD (1997), *Oslo Manual; Proposed Guidelines for Collecting and Interpreting Technological Innovation Data*, Paris: OECD.

OECD (2001), *Innovation and productivity in services*, OECD proceedings, Paris: OECD.

Park, Dong Bae (1999), *The Patterns and Trends of Technological Innovation in the Photonics and Optics in Korea* (In Korean), Seoul: Science and Technology Policy Institute.

Park, Sam Ock, and Choi, Ji-Sun (2005), IT service industries and the transformation of Seoul, in Daniels, P.W, Hutton, T.A, and Ho, K.O (eds) (2005) *Service industries and Asia-Pacific cities: new development trajectories*, London: Routledge.

Park, Sam Ock, and Nahm, Kee-Bom (1998), Spatial structure and inter-firm networks of technical and information producer services in Seoul, Korea, *Asia Pacific viewpoint*, **39** (2), 209–219.

Sirilli, Giorgio and Evangelista, Rinaldo (1998), Technological innovation in services and manufacutring: results from Italian surveys, *Research policy*, **27**, 881–899.

Um, Mi-Jeong and Choi, Ji-Sun (2004a), *Korean Innovation Survey 2003: Service sector* (in Korean), Seoul: Science and Technology Policy Institute.

Um, Mi-Jeong and Choi, Ji-Sun (2004b), *Analysis of the Technological Innovation in Service Sector with the Korean Innovation Survey Data in 2003* (in Korean), Seoul: Science and Technology Policy Institute.

Chapter 5

Location Patterns of Information Technology Services in Japan

Noboru Hayashi

Introduction

Information technologies have drastically changed the way in which contemporary firms in advanced countries do business. Without the computer-oriented systems controlling many companies' organization with information technologies, firms can't work well. Information technology services, a kind of producer service, help firms to carry out their daily business smoothly. Close interdependence between information technology services and firms defines the spatial integration of these economic activities. As information technology services are heavily dependent on information infrastructures, it seems that they provide services from remote areas using telecommunication means. But, in fact, they tend to be located close to firms agglomerated in the downtown of a city. This chapter shows how information technology services have developed in recent years, and are spreading in Japan. The analysis explores recent trends from two different spatial scales: inter-urban and intra-urban.

Location Trends of Information Technology Services

Change in the Number of Service Establishments

With the diffusion of the Internet from the middle of the 1990s, information technology services using the Internet have greatly increased in Japan. Information technology services are one of the industrial sectors called information technology industries (IT industries). Information technology industries include not only hardware production industries such as computer fabrication but also service industries that develop software and sell them in the market. It is useful to illustrate how these industries have been distributed in the country for a short time because they have a possibility to change the business style of contemporary Japanese firms. However, there are several problems to overcome when investigating these industries using the nation's statistical survey data. Designated statistics gathered by the state have long time intervals and are not suitable for comprehending the changing patterns that have occurred relatively recently. Therefore, it is not unusual to use the telephone directory data compiled by the affiliate company of NTT (Nippon Telegraph and Telephone Corporation) as a surrogate source. We can map the short-term changes in the location of information technology services by surveying these convenient data updated every half year.

In Japan, the Ministry of Land, Infrastructure and Transport (MLIT) has conducted surveys using the NTT's telephone directory data in order to chart the nation-scale trend of information technology services. According to the result of these surveys, three kinds of business type classified by the NTT's criteria: software-related services, data processing services and Internet-related services, can be grouped into an information technology industry. These are businesses providing various information services to consumers using information technology. It is noticeable that traditional service activities such as telephone services and telecommunication services are not included in this category. This is a burgeoning information service industry that has grown rapidly since the beginning of the diffusion of the Internet.

How has this service industry developed in recent years? To see this trend, we begin by looking at the change in the number of establishments every six months during the period from September 1999 to March 2002. The Internet-related services expanded especially quickly; increasing from 1,691 to 6,097 in one year since September 1999.

The growth continued further, and the number of establishments became 7,695 in the next year. Establishments newly added within six months amount to about 1,600. It is noticeable that the increase from September 1999 to March 2000 was so remarkable that the growth rate became 141.0 per cent. Compared to the Internet-related service, two other services: software-related service and data processing

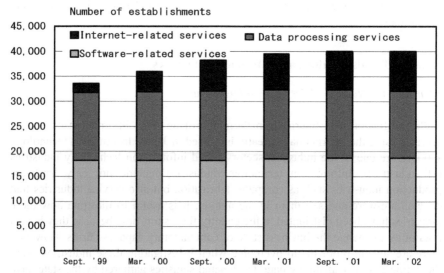

Figure 5.1   Trend in number of information technology establishments, Japan, 1999–2002

service showed relatively slow growth. There was no evident change in the number of establishments of software-related services over a two-year period. Only 424 establishments (from 18,141 to 18,565) were added in this period. In the case of data processing services, the number decreased from 13,637 to 13,506. In effect, the establishments of the information technology industry increased 20 per cent within two and a half years, but almost all the increase was attributed to the growth of Internet-related services. This change can be explained by the fact that the rapid advancement of innovation and diffusion related with the Internet media during this period in Japan inspired the development of the Internet-related services.

Thus, it can be said that the information technology industry has relied on Internet-related services for its expansion from the latter part of 1999 to early 2002. However, it should be noted that this trend was the result of the difference between absolute increase and decrease of the establishments. While there seems to be no change in the software-related service and the data processing service, the birth and death of establishments in fact happened. Accordingly, if we examine the patterns in the births and deaths of firms, we can clarify more clearly the trend of this service industry.

In the Internet-related services, which showed the largest relative increase, about 1,500 births were repeated every six months from September 1999 to March 2002. The number of births always outnumbered that of deaths, so the establishments continued to increase. However, the rate of birth began to decrease from September 2000, and the difference between the rate of birth and that of death diminished. This service sector seemingly continued to grow, but it should be recognized that the rate of death was about 15 per cent since March 2000. The rate of birth in the software-related services continued to outnumber that of death until September 2001, but the difference diminished within the next six months. The rate of birth was 15.9 per cent and that of death was 16.2 per cent in this period. Only this service sector increased in the rate of birth from March 2001. It is indicated from this that while the two other service sectors have gradually abated because of the nature of business, software-related services have still grown owing to the increasing demand for software development from firms.

The birth rate of data processing services began to decrease in March 2000, and continued to decrease afterwards. The rate of death outnumbered that of birth by March 2001, and the absolute number of establishments began to decrease from September 2000. It is very common that many firms enter and withdraw from the market in the information technology services. It is also common that firms change their businesses and develop new services. Some of firms dealing with data processing services switched from conventional or routine services to creative businesses; software-related services or Internet-related services, to pursue more profit. The trend of information technology services from September 1999 to March 2002 in Japan is characterized by the fact that while Internet-related services developed greatly in the early part of this period, software-related services expanded at the end of this period at the expense of the relative decline of data processing services.

Regional Location of Service Establishments and Trends in Opening and Closing

What is the regional distribution of information technology services in Japan?
According to the calculation of the share of service establishments by prefecture,
Tokyo has the largest share (30.5 per cent) and is followed by Osaka (9.4 per cent)
(Table 5.1). Kanagawa (6.1 per cent), Aichi (5.1 per cent), Fukuoka (4.1 per cent)
and Hokkaido (3.7 per cent) follow further in this order. This rank order is the
same as that of regional share by each kind of business. The total share of these
six prefectures amounts to almost 60 per cent in the software-related and data
processing services, but the total share is below 50 per cent in the Internet-related
services. This suggests that the establishments of the Internet-related services are
more widely dispersed than the other two services. These spatial patterns were stable
from September 1999 to March 2002. The share decreased by 0.7 points in Tokyo
and by 0.6 points in Kanagawa, respectively, but there was no clear change in the
other four prefectures.

It is, to a certain degree, predictable that the information technology industry will
be distributed in accordance with the size of prefecture on the basis of population

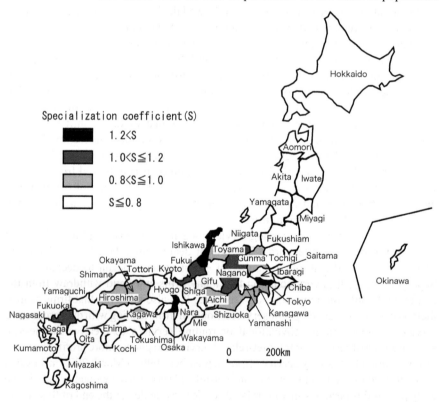

**Figure 5.2   Specialization coefficients for information technology services,
Japan**

Table 5.1 Number of establishments of information technology services by prefecture, Japan, March 2003

Prefecture	Software-related service	per cent	Data processing service	per cent	Internet-related service	per cent	Total	per cent
Tokyo	6,530	35.2	3,531	26.1	1,903	24.0	10,910	30.5
Osaka	1,712	9.2	1,239	9.2	590	7.4	3,349	9.4
Kanagawa	1,289	6.9	766	5.7	391	4.9	2,168	6.1
Aichi	888	4.8	769	5.7	373	4.7	1,824	5.1
Fukuoka	696	3.7	648	4.8	337	4.2	1,483	4.1
Hokkaido	572	3.1	591	4.4	322	4.1	1,329	3.7
Others	6,878	37.0	5,962	44.1	4,028	50.7	14,722	41.1
Total	18,565	100.0	13,506	100.0	7,944	100.0	35,785	100.0

Source: Dial Information Company of Nippon Telegraph and Telephone Corporation (2003)

or share of other economic activities. The specialization coefficient, which is calculated by dividing the share of service establishment by that of population of each prefecture, can be used to confirm this prediction. Among the top group by population, only three prefectures: Tokyo (3.21), Osaka (1.35), and Fukuoka (1.05) have a specialization coefficient greater than 1.0. These are the regions that have more service establishments than expected. In the lower group, Ishikawa (1.21), Fukui (1.08) and Nagano (1.08) are the prefectures that have specialization coefficients of more than 1.0.

The calculation of the specialization coefficient by each kind of business shows first, that for the software-related services, Tokyo (3.07), Osaka (1.33), Nagano (1.07) and Kanagawa (1.04) are the prefectures with more establishments than expected. Tokyo (2.75), Osaka (1.32), Ishikawa (1.20), Toyama (1.19), Nagano (1.11), Fukui (1.08), Aichi (1.03) and Miyagi (1.03) have relatively more data processing service establishments. The specialization coefficient for Tokyo of this service is lower than that of the software-related services and, in contrast, other prefectures show higher specialization coefficients than those of software-related services. This trend is strengthened for Internet-related services, because this service is spread more widely throughout the country. Tokyo's specialization coefficient is 2.52, and nine prefectures: Ishikawa (1.89), Fukui (1.79), Nagano (1.41), Toyama (1.34), Okayama (1.25), Hiroshima (1.24), Osaka (1.07), Fukuoka (1.07) and Saga (1.04) have specialization coefficients above 1.0. Summing up, it can be concluded that the information technology industry is mainly distributed in the prefectures with large metropolitan areas and the prefectures located in the central north (Hokuriku) and Nagano.

While the locational pattern of information technology industry in March 2002 is described above, what is the temporal change in the birth and death rate of establishments? If the three kinds of business are considered, it is clear that the number of birth and death is nearly equal in every prefecture. In Tokyo, for example, 757 firms opened between September 2001 and March 2002 and 717 firms closed during the same period. In Osaka, 226 firms entered into the market and 215 firms withdrew from it during the same period. Consideration of the change in each kind of business at national level reveals that there are regional differences. The birth rate of software-related services is generally higher in western Japan than in the eastern part of the country. The rate of birth of the data processing is higher in Hokkaido, Aomori, Kinki and northern Shikoku, while Tohoku, Kinki and Kyushu show a higher birth rate for Internet-related services.

Urban Hierarchy and Location of Information Technology Services

Tokyo has about 30 per cent of all the information technology service establishments, and the three largest metropolitan areas including Tokyo house more than 50 per cent of all firms. Fukuoka prefecture with Fukuoka as the largest city and Hokkaido prefecture with Sapporo also have a large share. From this, it is evident that information technology industries are mainly located in the metropolitan areas.

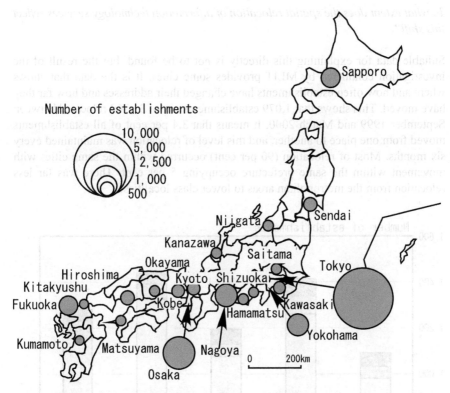

Figure 5.3   Agglomeration of information technology services, by city, Japan, 2002

An analysis of the proportion of these industries in the country's prominent cities reveals that Tokyo (only 23 ward areas) has a share of 22.7 per cent of the national total, the government-designated cities (12 cities in total), 25.0 per cent and other cities, 52.3 per cent. Tokyo's share of data processing services is 24.5 per cent and the government-designated cities 29.4 per cent of this service. This means that data processing services tend to be agglomerated in the larger cities. The share of software-related services among a smaller hierarchy of cities is very similar to that of the Internet-related service.

It has already been mentioned that the growth in the number of establishments began to slow down in every service sector after September 1999. This trend was common to all cities except Tokyo where the number continued to expand from September 1999 to September 2000. The government-designated cities in particular experienced a slow down allowing lower class cities to gain increased information technology services. This means that there has been a relative spatial disperse of service activities to local areas lower down the urban hierarchy.

To what extent does the spatial relocation of information technology services reflect this shift?

Suitable data for explaining this directly is not to be found, but the result of the investigation conducted by MLIT provides some clues. It is the data that shows where and how often establishments have changed their addresses and how far they have moved. This shows that 1,079 establishments out of 31,769 relocated between September 1999 and March 2000. It means that 3.4 per cent of all establishments moved from one place to another, and this level of relocation was maintained every six months. Most of relocation (90 per cent) occurred within the same cities with movement within the same prefecture occupying 5 per cent. There was far less relocation from the metropolitan areas to lower class localities.

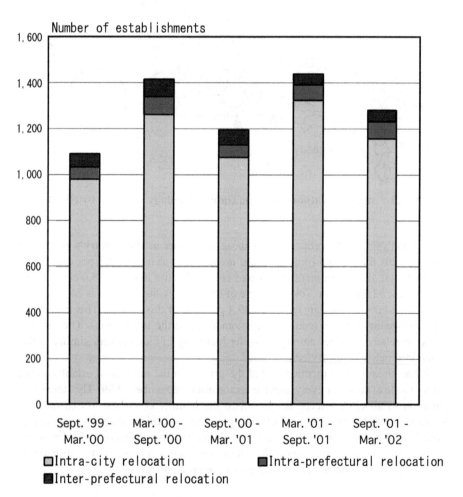

Figure 5.4   Regional relocation of establishments, Japan, 1999–2002

Location of Information Technology Services within the Metropolitan Areas

Location of Establishments in the Wards and around the Main Stations of Tokyo

Tokyo has about 30 per cent of all information technology services in Japan, and they are mainly concentrated in three central wards (Chiyoda, Chuo and Minato) and neighboring wards (Shibuya and Shinjuku). More than 800 establishments are located in each of the wards mentioned above, and these wards occupy 59.9 per cent of this metropolis. Every ward, except Chuo, has nearly the same share of establishments (12-13 per cent). It means that there is no clear regional differentiation in the location. Minato, Chiyoda and Shibuya continued to increase the shares after September 1999. From this, we can say that there seems to be an agglomeration trend within these wards. Shinjuku decreased its share from 12.5 per cent to 12.0 per cent within two and half years. There is some difference between Chuo's position and that of other main wards in that it increased its share from 8.4 per cent to 9.0 per cent. In any event, it is certain that there is a discrepancy between these main wards and other peripheral wards regarding the location of establishments. About 20 per cent of information technology services in Japan are grouped within the CBD and sub-centers of Tokyo.

There is no clear differentiation in the constitution of information technology services amongst the leading five wards of Tokyo. Overall, software-related service comprises about 50 per cent, data processing service about 30 per cent, and the Internet-related services about 20 per cent. The share of the Internet-related service is more than 20 per cent in Minato and Shibuya, and the share of data processing is

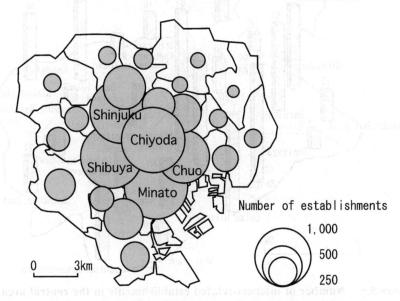

Figure 5.5   Location of information technology services in Tokyo, 2002

above average in Chuo and Chiyoda. Typically, during the year from March 2001 the establishments providing the Internet-related services increased greatly. In every ward of central Tokyo, the Internet-related service establishments increased by more than 10 per cent, Chiyoda especially showed a significant increase of 23.2 per cent. Software-related service establishments also increased in Chiyoda (5.4 per cent) while data processing establishments decreased in number. This was repeated in other main wards. Shinjuku and Minato especially lost establishments, possibly as a consequence of a shift from data processing to Internet-related services in this period.

The agglomeration of information technology services around Shibuya ward and Minato ward is called Bit valley or Japanese Silicon Valley.The reason why this area is called Bit Valley is that the Japanese place name, Shibuya, comprises two characters, Shibu and Ya, and Shibu means bitter in English. Bit, of course, is a unit word used in computers, so the word, Bit Valley, contains two kinds of meaning. There were many information and software firms in the early 1990s in these two wards. After the boom of the Internet-related services, the number of establishments increased sharply. Some firms shifted their activities away from data processing or software-related services to Internet-related services. The size of the Bit Valley agglomeration is almost the same as that of the Multimedia Gulch in San Francisco.

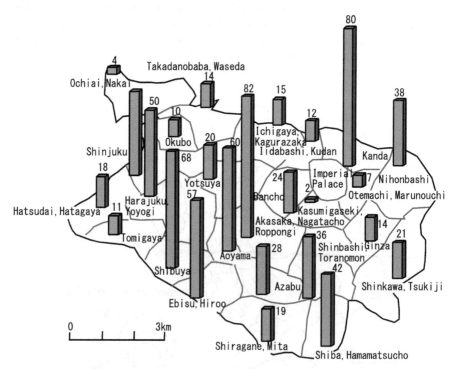

Figure 5.6   Number of internet-related establishments in the central area of Tokyo, 2000

Wards of central Tokyo

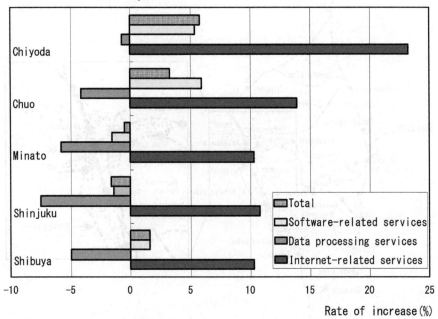

Figure 5.7   Rate if increase of information technology services, central Tokyo,
             March 2001–March 2002

Social and cultural amenities, artistic environment, and mass-communication medias have contributed to the formation of the regional agglomeration of information technology services there.

Using the results of the survey conducted by MLIT, we can show how many information technology service establishments are located within a 1km radius of the main railway stations. A location close to the main stations is advantageous for these kinds of services because external accessibility is regarded as important to them. When the number of establishments within 1km radius from the station is counted and arranged in order, Akihabara (Chiyoda ward) emerges as the station with the largest agglomeration of establishments (720) in March 2002. Shibuya (Shibuya ward) is second (490), followed by Tochomae (Shinjuku ward) (410), Kayabacho (Chuo ward) (420), and Ikebukuro (Toshima ward) (390). These five principal station districts contain 59.7 per cent of all the information technology establishments in Tokyo. This confirms that good accessibility to the main station is a crucial locational factor for these activities. Akihabara is only the place where the number of establishments continued to increase during the survey period.

When the observation period is limited to the six months from September 2001 to March 2002, it is clear that establishments increased mainly in the places around Akihabara, Shibuya, and Tochomae stations.

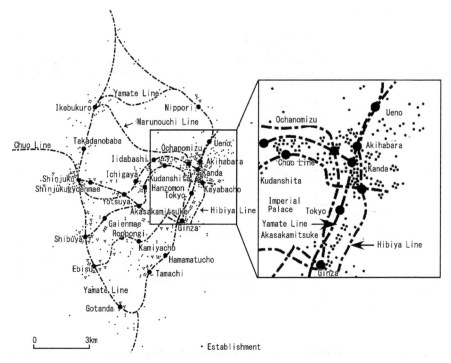

**Figure 5.8   Location of information technology services opened in the central
area of Tokyo, September 2001–March 2002**

Among them, Akihabara and its neighbouring stations, Kanda and Ochanomizu,
are places that attracted numerous establishments. Newly located establishments are
more often found in these places than in Shibuya and Akasaka, which are called Bit
valley. In Tokyo, where the railroad density is comprehensively high, establishments
increased not only near the main stations but also along the arterial railroads such as
the Chuo-line, Yamate-line, Hanzomon-line and Hibiya-line.

**Location of Establishments in the Wards and around the Main Stations of
Osaka and Nagoya**

Osaka is the second largest prefecture for the location of information technology
services, with the majority concentrated in the city of Osaka. It had 2,630
establishments in March 2002. There are four main wards with more than 370
establishments, they are Chuo (796), Kita (597), Yodogawa (389) and Nishi (375).
They occupied 80.2 per cent of all ITS firms in the city of Osaka. In Chuo ward,
the largest agglomeration, the share of software-related service was 51.4 per cent,
and data processing 33.4 per cent. The share of the Internet-related services was
relatively low. The composition of information technology services in Chuo was

similar to that of Kita, the second largest agglomeration. The share of the Internet-related services was also low in Yodogawa. More than 60 per cent of establishments in this ward were involved with data processing.

In Osaka, like Tokyo, the data calculated for defining the locational agglomeration around the main stations are available. Shinsaibashi, Shin-Osaka, Minamimorimachi and Umeda are the main stations around which many establishments were located. Among them, Shinsaibashi was the largest place (420) followed by Shin-Osaka (370) and Minamimorimachi (360). Three main agglomerations could be found: Shinsaibashi, Minamimorimachi=Umeda and Shin-Osaka, but the density of establishments was lower than that in Tokyo.

Nagoya is the third largest agglomeration for the establishment of information technology services, but its size is only half that of Osaka. The number of establishments by ward is as follows: Naka (516), Nakamura (202) and Higashi (110). These central wards occupy 65.8 per cent of all ITS in the city of Nagoya. In Naka, the largest agglomeration, 50.5 per cent is software-related service, and 36.0 per cent is data processing service. Nakamura, the second largest agglomeration, has a similar structure to Naka. Like the central area of Osaka, the share of the Internet-related service is low. The locational density of Nagoya is also lower than Osaka, and there are only two instances where the number of establishments within 1 km radius from the main station is more than 100 (Figure 5.9).

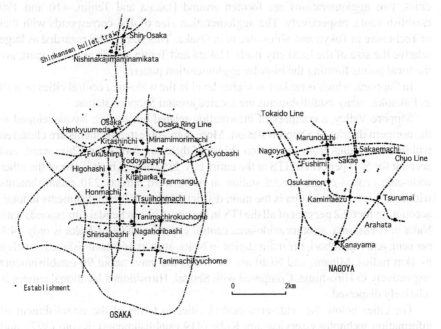

Figure 5.9   **Location of information technology services newly opened in central areas of Osaka and Nagoya, September 2001–March 2002**

Location of Establishments in the Lower Class Cities

Where are the Establishments of Information Technology Services Located in the Lower Class Cities?

Yokohama, which is the largest after Nagoya on the basis of number of establishments (1,143), is followed by Fukuoka (1,011), Sapporo (908), Sendai (520), and Hiroshima (509). This means that, with the exception of Yokohama, so-called wide area central cities: Sapporo, Sendai, Hiroshima, and Fukuoka have more than 500 establishments, respectively. This is not out of line with the urban hierarchical structure in so far as the functional structure of Sapporo and Fukuoka are roughly the same, and Sendai and Hiroshima are also the same in terms of the agglomeration of establishments. Summing up, the locational patterns of information technology services resemble the features of the urban hierarchical system depicted by the distribution patterns of wholesale institutions or headquarter offices.

Yokohama, Kannai, and Shin-Yokohama are the main stations in Yokohama and each has more than 100 establishments located within a 1km radius. Yokohama and Shin-Yokohama are 5km apart, so the agglomerations are regarded as distinct entities. Kannai, 2.7km southeast to Yokohama and the nearest station to the new civic center has attracted a lot of establishments because of its locational convenience. In Fukuoka, which is regarded as ranking higher rank amongst the wide-area central cities, two agglomerations are formed around Hakata and Tenjin, 410 and 190 establishments, respectively. The agglomeration size of 410 corresponds with that of Tochomae in Tokyo and Shinsaibashi in Osaka. This size can be regarded as large relative the size of the local city itself. Hakata and Tenjin, which are 2km apart, are the focal points forming the bi-polar agglomeration pattern.

In Sapporo, which is ranked as higher level in the wide-area central cities as well as Fukuoka, many establishments are located around Sapporo station.

Sapporo Valley, a complex of information related businesses, has developed in the northern district of Sapporo station. More than 200 establishments are clustered within 1 km radius from the station. Most of them are formed in Chuo ward, and account for 50.7 per cent of ITS in the entire city. In Sendai, which is one of the other wide-area central cities, Sendai station in Aoba ward attracts 230 establishments within its 1 km radius. Aoba is the main district that has 360 establishments in total, accounting for 69.2 per cent of all the ITS in the city. Aoba in Sendai corresponds with Naka in Hiroshima, another wide-area central city. The share of Naka is only 44.8 per cent, and Hachobori, the main station in Naka, attracts 160 establishments within its 1km radius. Minami and Nishi are the wards that have about 90 establishments respectively in Hiroshima. Compared with Sendai, Hiroshima's locational pattern is relatively dispersed.

The cities below the wide-area central cities in terms of the establishment of information technology services are Kobe (419 establishments), Kyoto (397), and Kawasaki (389). Although Saitama (252), Kita-Kyushu (230) and Chiba (195) are government-designated cities, their ITS agglomerations are smaller than those of

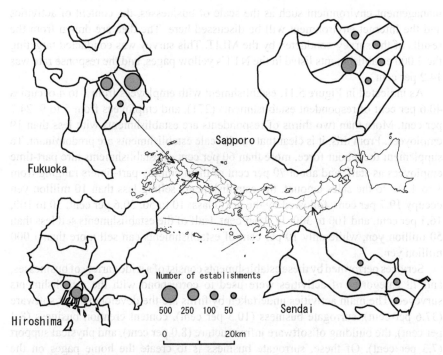

Figure 5.10 Location of information technology services in the regional core cities, Japan, March 2003

Okayama (284), Kanazawa (283), Niigata (259) and Hamamatsu (256), which are not government-designated cities. In Kobe, Chuo ward accounts for 56.6 per cent of the city total. Shimogyo ward in Kyoto, and Kawasaki ward in Kawasaki have smaller share than that of Chuo ward in Kobe, but the agglomeration sizes are larger than Kobe. This means that the locational pattern is more dispersed in Kyoto and Kawasaki than in Kobe. Kokura ward in Kita-Kyushu has 47.4 per cent and Naka ward in Chiba has 48.2 per cent respectively. In Kanazawa, Niigata and Hamamatsu, the main stations attract a lot of establishments. There are 100 establishments within a 1km radius from Niigata station, and 60 establishments exist in the same area of Hamamatsu station.

Activities of Information Technology Services and their Changes

Nature and Environment of Information Technology Services

As indicated in the previous section, ITS are located in the metropolises centering on Tokyo and Osaka. They are mainly located around the central stations and the CBDs. What are the environmental conditions for ITS like? How are software-related services, data processing services and Internet-related services being provided? The management environment such as the scale of businesses, the content of activities and the Internet environment will be discussed here. The data are drawn from the results of the survey conducted by the MLIT. This survey was conducted targeting the 2,000 establishments listed in the NTT's yellow pages, and the response rate was 14.2 per cent.

As indicated in Figure 5.11, establishment with employees from 1 to 4 occupies 40.6 per cent of respondent establishments (271), and employees from 5 to 9, 24.7 per cent. More than two thirds of respondents are establishments with less than 19 employees. From this, it is clear that small-scale establishments are predominant. To supplement the labour force, more than 60 per cent of establishments hire part-time employees as staff, and about 70 per cent of them employ part-timers ranging from 1 to 4. As to the sales amount, the establishments selling less than 10 million yen occupy 19.7 per cent, followed by the next class: 10 to 50, 30.6 per cent, 50 to 100, 16.1 per cent, and 100 to 300, 11.9 per cent. Half of the establishments sell less than 50 million yen, while only 10 per cent of establishments can sell more than 1,000 million yen.

Services performed by these establishments consist of a wide variety of businesses, and 13 categories of activities were used to correspond with the establishments surveyed. The main activities undertaken by firms are the development of software (37.6 per cent), surrogate business (10.3 per cent), content creation business (8.4 per cent), the building of software infrastructure (8.0 per cent), and physical support (7.3 per cent). Of these, surrogate business is to create the home pages on the Internet, input data, and the management of system servers. The building of software

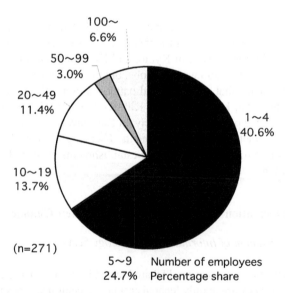

Figure 5.11 Size distribution of information technology service firms established April–September 2001, Japan

infrastructure is part of planning information systems and networks. Physical support includes consulting, manpower dispatching and educational training. Most of the establishments conduct several different activities, and the proportion of surrogate business is larger than that of software infrastructure.

As 79.1 per cent of respondent establishments are headquarters offices, it is clear that this kind of business is performed independently. Regional offices account for only 9.5 per cent, and branch offices, 8.8 per cent. Small scale and degree of independence of establishments is related with the narrowness of their activities. Most of these activities are carried out in the small rooms of office buildings. The establishments with office space below 50m^2 occupy about half, and 73.3 per cent of the establishments have office space below 70m^2. Around 85 per cent of establishments use rental properties, and 40.5 per cent of them pay 10,000 yen as rental expense for unit area (3.3m^2) and 31.1 per cent, 15,000 to 20,000 yen. Good accessibility to the Internet environment is very important for ITS. The most popular way for connecting with the Internet is ADSL (Asymmetric Digital Subscriber Line), which is used by 38.3 per cent of firms. ISDN (Integrated Services Digital Network) at 29.4 per cent is the next in the list followed by dial-up service (ISDN and analogue) (17.8 per cent) and business track (17.5 per cent). Business track and dial-up services have declined in popularity as access to the Internet by using CATV (9.3 per cent) and optical fibre (4.5 per cent) have become popular.

Location Factors for Information Technology Services

As for other types of business establishment, the minimization of production costs and the maximization of sales volume are important factors to be considered when exploring the location of information technology services. To minimize production costs, some caution over labor cost, rental charges for floor space, procurement costs, and transportation and communications cost is needed. On the other hand, optimizing accessibility to clients, innovation in the services provided, and reasonable prices/ fees are necessary to achieve an increase in the volume of sales. As the difference between the sales amount and the cost is profit, the greater the difference, the more successful the business.

In the survey conducted by the MLIT, 35 location factors were included. Respondents were asked to indicate for each factor whether it was a special consideration, a consideration, or not a consideration. Whether the factors were a special consideration or a consideration, are focused on here. It is possible to classify the 35 factors into seven groups: the availability of office space, the accessibility of transportation, the convenience of access to the activities, proximity to work place and residence, the quality of environment, the ease of acquiring human resources and technologies, and others. These can be regrouped into just two categories: the minimization of cost and the maximization of sales.

For ITS the factors associated with proximity are regarded as important; accessibility to their clients (80.9 per cent), accessibility to the nearest station

(77.6 per cent), accessibility by train (73.0 per cent), proximity to the residence of corporate managers (56.1 per cent), proximity of to the residence of employees (55.8 per cent), and accessibility to subcontract companies (55.2 per cent). The proximity to shareholders and headquarters gain 39.5 per cent and 35.9 per cent, respectively. Accessibility to highway (53.5 per cent) is lower than that of railroads, but is higher than airports. Commuting costs, travel expenses, and selling expenditure, which are measured by transportation and communication cost, are related with either the control of production cost or the expansion of sales volume. In any event, it is evident that accessibility and proximity are regarded as especially important for the location of ITS establishments.

The level of rent for leased space was regarded as the most essential condition, 93.3 per cent chose this item. Facilities and equipments are indispensable conditions, and adequate building (70.8 per cent), good environment for telecommunication (69.5 per cent), regional infrastructure for telecommunication (68.5 per cent) are included in this category. The familiarity of the city and region (62.9 per cent) and the regional image (58.5 per cent) are related with the reputation of location, and may contribute to the increase of sales amount with the credit and public image. The existence of many nearby client firms (53.0 per cent), the substantial provision of business assistance close to the office (40.0 per cent), and the affluent subcontract firms at a short distance (36.8 per cent) again indicate the importance of proximity. Summing up, it is clear that establishments place greater emphasis on not only facilities and equipment but also on the environmental conditions for their location.

Locations of Subcontract Firms and Clients and the Extent of the ITS Labor Market

When information technology services are produced and supplied, proximity is very important at the stage of acquiring information as material as well as selling processed services. But, is it really possible for the firms to be located close to the clients and to their subcontracts? In the survey conducted by the MLIT, the addresses of subcontract or companies and clients were identified. It emerges that the share of subcontract companies in the same local municipality as their ITS partners is the largest (Figure 5.12), at 50.4 per cent, followed by the same prefecture (42.1 per cent), the ward area of Tokyo (40.3 per cent), and the city of Osaka (12.0 per cent). This suggests that firms purchase material input, information and the financial service that they require from the neighborhood.

Correspondingly, the main clients of firms tend to be found in the same municipality (52.6 per cent). The equivalent proportions for the same prefecture are 48.6 per cent, the ward area of Tokyo (41.9 per cent) and the city of Osaka (15.0 per cent). The pattern is almost the same as that for subcontract companies, and it means that interactions occur within a relatively compact area. This somewhat contradicts the image that ITS are exchanged widely, regardless of distance. The emphasis on proximity within a relatively small area helps to explain why information technology

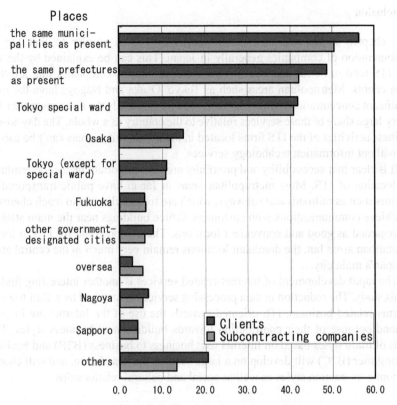

Figure 5.12 Location of subcontracting companies and clients

services tend to occur in agglomerations. While it is now not difficult to conquer time distance, good accessibility to the site of procurement and distribution is still important for location.

It is not only ITS but also other services that need high quality human resources as their indispensable condition of activities. However, as high quality staff are valuable, it is critical to keep a balance between cost and benefits. As to the question as to where staff were mainly recruited, most of the firms indicated that they are hired from within the local area. The main ways of recruitment are as follow: through a personal connection (39.5 per cent), from the employment agency (37.2 per cent), on Internet sites (32.9 per cent) and by recruitment articles in newspapers and magazines. It is interesting to know that the percentage of permanent hiring after part-time working and head hunting from shareholders alike are more than 13 per cent, respectively. The percentages of direct recruitment from the universities and job fairs sponsored by the municipalities are lower, suggesting that institutional ways of recruitment do not work well.

Conclusion

This chapter shows that the spatial pattern of ITS is similar to that for the agglomeration of companies generally in Japan. This can be explained by the fact that ITS tend to be located close to the concentration of companies that are their main clients. Metropolitan areas such as Tokyo, Osaka and Nagoya have the most significant concentration of these services. Tokyo, especially the downtown area, has a very large share of these services relative to the country as a whole. The day-to-day business activities of the ITS firms located in the metropolitan areas can't be carried out without information technology services.

It is clear that accessibility and proximity are very important factors determining the location of ITS. Most metropolitan areas in Japan have public transportation systems such as railroads and subways, which are frequently used to reach clients or to achieve communications with customers. Office buildings near the main stations are regarded as good and convenient locations. There is potential for ITS to locate in suburban areas but, the dominant locations remain very much in the central areas of Japan's main city.

The rapid development of Internet-related services is another interesting finding in this study. The reduction in data processing services is matched by a shift towards Internet-related business. ITS oriented towards the use of the Internet are in great demand because of their contribution towards building new business styles. Two kinds of trade styles based on Internet use: business to business (B2B) and business to consumer (B2C) will develop on a large scale in the near future, and will change economic interaction styles as well as social and cultural relationships.

References

Dial Information Company of Nippon Telegraph and Telephone Corporation (2003), *Town Pages*, Tokyo and Osaka: East and West Nippon Telegraph and Telephone Corporation.

Ministry of Land, Infrastructure and Transport (2002), *Research Report on Software-related Information Technology Industry*, Tokyo: Department of Printing of Ministry of Finance.

PART II
External Regulation of Services
within Value Chains

Chapter 6

State, Market and the Growth of Service Industries in Metropolitan Guangzhou

Fiona F. Yang[1] and George C.S. Lin[1]

Introduction

Over the last half-century, advanced economies have experienced a 'quiet revolution' through which services have replaced manufacturing as the predominant production activities (Channon, 1978, Inman, 1985). Since service industries have tended to be spatially centralized, reflecting the influence of urban and agglomeration economies, the growth of services has important implications for urban transformation and the reformation of urban hierarchies. The significance of service industries to economic and societal transition has been well documented in a large body of literature (Fisher, 1939, Clark, 1940, Bell, 1973, Daniels, 1982, Allen, 1988, Marshall et al., 1987, Marshall and Wood, 1995, Tickell, 1999, 2001, Coffey, 2000). In the current era of globalization, the interrelationship between service industries and urban change has been evaluated at various geographical scales.

First, from the global perspective, the 'global shift' of industrial production is believed to have given rise to the emergence of 'world cities' (Friedmann and Wolff, 1982, Friedmann, 1986), 'global cities' (Sassen, 2000, 2001), and 'global city-regions' (Scott, 2001). While manufacturing activities have dispersed to those cities and regions with lower production costs, control, management and innovative functions have tended to be concentrated in large metropolitan areas with advanced infrastructure and communication facilities (Moulaert et al., 1995). These metropolises are 'strategic' loci in organizing international expansion of capital, directing global industrial production, and producing specialized services (Sassen, 2000). As the 'base points' in the new spatial division of labor, global cities or city-regions preoccupied with producer services are influential in the repositioning of the world city systems.

Second, the scalar reshuffling of power both upward and downward has led to the formation of 'boosterist' coalition that drives the urban 'growth machine' to capture the exchange-value of the city property and maximize financial gains from their assets (Logan and Molotch, 1984, Lin, 2002: 299). Resulting from the penetration of global forces or the changing geography of production, many Western cities have suffered

1   Department of Geography, University of Hong Kong, Hong Kong.

from the process of 'deindustrialization' in the early 1970s (Kirby, 1986). In tandem with the regional economic restructuring, there has been a broad political transition from municipal managerialism to urban entrepreneurialism and from government to governance in the era of intensified global competition (Harvey, 1989; Hall and Hubbard, 1998). In order to lure footloose global capital for local economic growth, the 'new urban politics' (Cox, 1993) has constructed place-specific advantages and built up attractive city images. Such 'projects' as land property development, establishment of industrial districts or high-tech parks, city center re-formation, and waterfront redevelopment have been carried out, which have extended well beyond the sphere of production toward consumption and engendered observable new change in the urban landscape.

Third, when attention is shifted from the Atlantic core to the Asia Pacific, the growth of service industries has been regarded as 'a concomitant element of advanced production systems, national modernization programmes, and globalization strategies of central and regional governments' (Hutton, 2004: 22). As is well known, the 'East Asia economic miracles' over the last half century have been underpinned primarily by the process of rapid industrialization in the region. While industrialization programmes may vary from place to place, manufacturing-led policies have maintained a strong position in the national or regional developmental agendas. In recent years, in addition to the robust growth of manufacturing activities, service industries have received increasing attention with respect to both sector-based policies and general development strategies (Hutton, 2005). It is argued that the accelerated tertiarization processes in the NICs since the 1980s have been the reflection of policy efforts to support and promote urban economic upgrading.

Rapid expansion and development of service industries have also been evident in China since the economic reforms initiated in 1978. Recent economic restructuring and spatial transformation in China has demonstrated a distinctive departure from the socialist patterns characterized by centralized planning, low-level urbanization, and city-based industrialization. While the tertiarization process in China is by no means 'post-industrial' (Lin, 2005), it is believed that industrial-deterministic discourse of urbanization may no longer be adequate to enlighten the complex patterns and processes of urban transformation (Lin, 2004, 2005, Yang, 2004). How are we to understand the growth of service industries and its impact on urban development in the Chinese context? When the extant theories are used to explain the case of China, however, there exist apparent discrepancies and inconsistencies since the growth of services in China is embedded in a politico-economic context different from that of capitalist advanced economies.

A problematic assumption upon which the world/global thesis is based is that the nation-state is declining in importance as transnational capital becomes increasingly 'placeless'. As Sassen (2000: 56) argues, the formation of a 'transnational urban system' entails '... the reduced role of the government in the regulation of international economic activity and the corresponding ascendance of other institutional arenas, notably global markets and corporate headquarters'. The powerful Party-state in China has undoubtedly limited the applicability of the thesis (Ma, 2002). The neo-liberal discourse which highlights the role of urban governance is important to conceptualize the uneven growth of service industries in China. It has witnessed a transition from centralized control to decentralized decision-making since the 1980s which has given rise to an empowered local government and changing urban governance (Oi, 1992, 1995, Lin, 2002, Wu, 2002). However, the intensive intervention of the state in the economy and the great influence of the nation-state in the Chinese context have significantly undermined the explanatory power of the neo-liberal discourse. With a similar culture and development trajectory, the Asia-Pacific experiences appear to be highly relevant to analyze the growth of services in China. A closer scrutiny reveals, however, that the developmental state fails to 'accommodate local aspects of developmentalism and the local efforts for development from the local states' (Xia, 2000: 26). A critical evaluation of the existing literature against the Chinese experience would suggest that the transformative path of a city toward a service economy has been significantly shaped by the way in which state and market interact. As Bryson and Daniels (1998: xix) stress, we should 'take into consideration historic, social, political or cultural differences between countries' to understand the growth of services.

This chapter is an attempt to explore a more inclusive theoretical explanation for the growth of service industries in China. We argue that the process of tertiarization and urban transformation cannot be fully explained without a comprehensive understanding of the state-market relation which is temporally and spatially contingent. This chapter examines the role of state-market interaction in the growth of service industries in the Chinese context, taking Guangzhou, one of the most rapidly expanding and 'globalizing' metropolitan areas in China as a case study. The next section contextualizes the changing state-market relation and the growth of service industries over the past decades. This is followed by a study of the growth of services in Guangzhou, and an analysis of the role of state-market relation in urban transformation of the metropolis. The brief conclusion highlights important findings of the study and raises a number of theoretical questions for further investigations.

State, Market and the Development of Service Industries in the Chinese Context

As O'Neill (1997) notes, the economy is necessarily a combination of markets, state action and state regulations. The way in which state and market interact therefore shapes the development trajectory of the economy. Over the last half century, China has been undergoing significant institutional and ideological changes, which has given rise to its peculiar politico-economic context and the unique process of economic restructuring.

It is commonly acknowledged that the Chinese socialist state adopted a planned economic system in Mao's era. The urge to limit and control the functioning of the market stems from the Marxist ideology that market and private property inherently produce inequality and alienation (Block, 1994). In order to eliminate injustice, the socialist state managed to establish a centralized framework for mobilizing and allocating resources. Free market was not allowed to exist in 1953 and market was limited only for 'commodities'; no markets existed for labor and natural resources since they were 'state-owned' (Hsu, 1985). The negative perception of market engendered a hostile attitude toward service activities which were believed to help expand the market. The general assessment of trade and commerce was that 'no new wealth is created for society, so that the surplus value (profit) that is appropriates is actually merely a portion of the surplus value already yielded by productive (industrial) capital' (Solinger, 1983: 195). Regarded as 'non-productive' and exploitative, the development of services was restricted. Following the Soviet example, the Maoist regime gave priority to the development of heavy industry, concentrating resources inputs in the industrial sector by 'squeezing' those in the agricultural and service sectors. For ideological and strategic reasons, socialist cities were categorized into 'producing' or 'consuming' types, and those classified as 'consuming' had to be structurally transformed into 'producing' ones (Lo, 1994).

The Chinese state has, since the 1980s, initiated a series of economic reforms to relax and liberalize the market. The market economy assumed a legitimate status in China in October 1984 when the 'CCP Central Committee's Decision on Economic System Reform' stated that the socialist economy is a planned commodity economy based on public ownership (Hsu, 1985). In contrast to the Maoist regime, market in post-Mao China has been considered 'an economic mechanism that can produce some increase in efficiency' (Smart, 1997). The scope of the market has been progressively expanded so that 'a host of structural weakness that had led to recurrent inefficiencies and stagnation in the operation of China's socialist planned economy' can be overcome (Solinger, 1993: 107). The changing ideology of the market in the Chinese context has resulted in a fundamental alteration in the attitude toward service development. In 1985 the State Statistical Bureau set up the statistical system for the service sector (or 'tertiary industry' in the Chinese official definition), and explicitly referred to the sector as the industry that is 'not included in primary and secondary industry, and provide services of various kinds for production and consumption' (CSSB, 2001a: 84). Rather than arbitrary suppression, the service sector has been actively promoted since it appears to be not only an important source to absorb labor , but also an indispensable means to make available a wide range of consumer durables to raise the living standard of the general population. Supporting actions such as increasing investment in the sector, and encouraging market competition by creating multiple channels of service provision have subsequently been taken by the state to accelerate the expansion of services in the reform era. The growth of services has been one of the key forces driving the dramatic urban transformation, particularly metropolitan development in China in the recent decade. However, since the transition of the political economy from the Maoist to post-Mao regime has occurred in an *evolutionary* rather than a *revolutionary* manner (Lin, 1999), the state-market relation is peppered with 'Chinese characteristics':

'the communist party is still in power; distortions of economic factors are abundant; property rights are not yet clearly delineated; no large-scale privatization of state enterprises has been declared and the process of change is often marked by 'two steps forward and one step backward" (Zhang, 2000: 1).

As a consequence, the nature of services and the process of tertiarization in China stand in stark contrast with those found in advanced economies (Lin, 2005).

Service Industries and Urban Transformation in the Guangzhou Metropolis

Originally developed as a mercantile city, trading and service activities occupied an essential position in Guangzhou's urban economy even after the city suffered from the war in the 1940s. When the Communists took over Guangzhou on October 14 1949, the tertiary sector accounted for 47.06 per cent of GDP whereas the secondary sector contributed 42.98 per cent and the primary sector 9.96 per cent. The leading role of the tertiary sector was much more apparent as far as job provision was concerned. The tertiary sector provided jobs for 43.83 per cent of the labor forces while the secondary sector accommodated 20.59 per cent and the primary sector 35.58 per cent (see Table 6.1). The mercantile nature of Guangzhou, however, was condemned under the new regime for ideological and strategic reasons and had to undergo economic transformation.

In view of Guangzhou's pre-1978 comprehensive city plans (for detailed goals and objectives of the plans, see Xu and Ng, 1998: 44, Table 1), the inclination of the municipal government to transform Guangzhou from a 'city of consumption' into a 'city of production' was evident. Great efforts had been taken to fulfill this transformation. First, total fixed asset capital investment was redistributed and concentrated in the industrial sector at the expense of services since the First Five

Table 6.1   GDP and employment of the city of Guangzhou, 1949–2003 (%)

	1949	1965	1978	1985	1990	1995	2000	2003
GDP	100.00	100.00	100.00	100.00	100.00	100.00	100.00	100.00
Primary	9.96	4.79	3.84	3.99	3.54	2.67	3.03	2.34
Secondary	42.98	63.15	65.08	53.40	41.87	45.04	41.73	41.14
Tertiary	47.06	32.06	31.08	42.61	54.59	52.29	55.25	56.53
Employment	100.00	100.00	100.00	100.00	100.00	100.00	100.00	100.00
Primary	35.58	22.61	21.58	13.44	11.78	10.06	13.41	13.22
Secondary	20.59	39.39	45.18	45.37	42.56	39.97	41.64	38.44
Tertiary	43.83	38.00	33.24	41.19	45.66	49.97	44.95	48.34

Sources: Guangzhou Statistical Bureau (1999a: 209, 247); Guangzhou Statistical Bureau (2001: 29) Guangzhou Statistical Bureau (2004: 24, 80).

110   *Services and Economic Development in the Asia-Pacific*

Year Plan (1953–1957). By the year 1962, investment in the tertiary sector had substantially dropped from 74 per cent to only 33 per cent, whereas the secondary sector increased its share dramatically from 25 per cent to 62 per cent in ten years (Lin, 2005: 286). Second, a state-run commercial system replaced the natural evolving one, which further eroded the trading and commercial base of Guangzhou's urban economy. With the introduction of the system of 'unified procurement and distribution' in the late 1950s, a free market ceased to exist in China. The urban economy of Guangzhou was assaulted, trading and service activities shrinking significantly.

The result of the transformation policy to restrict consumption and encourage industrialization was a drastic reduction of GDP as well as labor force in the tertiary sector, although Guangzhou has for many years been short of industrial mineral resources. Table 6.1 shows the contribution of each sector to the economy of Guangzhou. The opposite trends of the secondary and tertiary sectors were obvious. Between 1949 and 1978, the secondary sector increased by 22.10 per cent in GDP and by 24.57 per cent in employment. The tertiary sector, in sharp contrast, decreased by 15.98 per cent and by 10.59 per cent correspondingly. Within the anti-free market context in Mao's China, like other Chinese cities, development of services in Guangzhou was depressed.

The implementation of economic reform since 1978 and Open Door policy since 1979 has set Guangzhou on a new path of economic restructuring and urban development. Because of their coastal frontier location, Guangzhou as well as the Pearl River Delta was considered 'insecure' in the Maoist regime and was subordinate to the interior with respect to the state budgetary allocation. This geographical location has turned into a unique advantage for the development of the city and the Delta region which are both outward oriented. Guangdong and Fujian provinces were allowed to practice 'special policies' in 1979 to stimulate foreign investment. Following the establishment of four Special Economic Zones, Guangzhou was designated as one of the 14 'open coastal cities' in May 1984. With recent economic development, Guangzhou has revitalized its former role as a commercial and trading port (Lo, 1994).

The intrusion of the new forces of globalization and marketization has significantly facilitated the reformation of the urban economy in Guangzhou. Statistical data indicate that the share of the tertiary sector enjoyed a dramatic growth from 31.08 to 56.53 per cent in GDP and from 33.24 to 48.34 per cent in employment during 1978 to 2003. The corresponding proportion of the secondary sector, in contrast, decreased from 65.08 to 41.14 per cent and from 45.18 to 38.44 per cent. As a consequence, the tertiary sector has exceeded the secondary sector to be the leading sector in Guangzhou in terms of both GDP and employment by 1989 (Yang, 2004). A closer scrutiny, moreover, unravels a sudden upsurge of the tertiary sector in the urban economy of Guangzhou in the early 1990s (GZSB, 2004: 24, 80).

The year 1992 is commonly admitted as a landmark in China's market-oriented reforms when the fourteenth National People's Congress initiated a large-scale reshuffling of the state-owned enterprises. Realizing that to deepen the reforms of the national economy might give rise to massive lay-offs (Putterman and Dong, 2000), the 'Resolution to Speed up the Development of the Tertiary Industry' was issued in the same year. The 'Resolution' emphasized the tertiary sector as an effective way to cultivate economic growth and to provide job opportunities for urban youths, factory layoffs, and rural migrants (Compilation Committee of China's Tertiary Industry Almanac, 1993). A series of approaches to speed up the development of services were adopted, including introducing a competitive mechanism, reforming enterprise management, promoting financial and taxation support, simplifying establishment procedures for new service firms, and formulating industrial regulation. The growth of service industries has since then become 'not simply a market-driven economic affair but one on the political agenda with important implications for urban manageability and social stability' (Lin, 2005: 291). The role of service industries in the economy has attracted increasing attention from the Chinese government in the successive fifteenth and sixteenth National People's Congress. In January 2002 the State Council and National Planning Committee jointly issued the 'Policies and Approaches to Facilitate the Development of the Tertiary Industry during the Tenth Five-Year Plan Period (2001–2005)'. Based on the 1992 'Resolution', the documentation ten years later further encourages service development by significantly expanding the scope of foreign capital in the tertiary sector (*People's Republic of China Yearbook*, 2002). In the Chinese context, the growth of services appears to be a process driven not so much by free market forces, but more by the state for political and social considerations (Gong, 2002, Lin, 2003).

Municipal fixed asset capital investment and city planning are two major instruments the Chinese state used to regulate economic activities and arrange land utilization. The new emphasis on service industries in national developmental agendas has greatly influenced the allocation of state investment in the economic sectors in Guangzhou. The municipal government has dramatically augmented the investment in the tertiary sector since the 1990s. Between 1996 and 2000, the capital invested in the tertiary sector was 81 billion yuan, which accounted for 84 per cent of total fixed asset capital investment in the city (GZSB, 2001: 125). The inclination of the city government to accelerate the growth of services is more evident when the urban planning of Guangzhou is analyzed. Table 6.2 compares current land use structure in 2001 with the planned one by the year 2010 in Guangzhou's most recent master plan. The land allocated for industrial use accounted for less than 20 per cent of the planned area, which represented a reduction of over 5 per cent from the current standing. The land for urban services, in contrast, was designated to expand from 32.57 to 44.71 per cent within ten years.

In addition to internal impetus, with the post-socialist change in state-market interaction, foreign investment has been one of the new forces stimulating the expansion of service industries in China. The important role foreign investment played in the process of industrialization has been stressed by many China analysts (e.g. Sit and Yang, 1997), but its contribution to the process of economic tertiarization has been less evaluated. Figure 6.1 demonstrates the changing sectoral distribution of utilized foreign investment in Guangzhou municipality (*shi*) over the period

Table 6.2   Land use structure of "2001–2010 Guangzhou Master Plan"

Land Type	Current		Planning	
	Area (sq. km)	%	Area (sq. km)	%
Total	557.61	100.00	755.24	100.00
Residential	179.92	32.27	202.00	26.13
Public Facilities	62.64	11.23	83.30	11.02
Industrial	139.81	25.07	149.83	19.82
Storage	13.06	2.34	20.97	2.77
Transportation	28.57	5.12	33.92	4.49
Road & Square	60.21	10.80	113.68	15.04
Urban Facilities	11.6	2.08	16.74	2.21
Greenland	47.18	8.46	124.25	16.44
Others	14.62	2.62	10.55	1.40

Note: The land for urban services includes land use for public facilities, road and squares, urban facilities and Greenland.
Source: Guangzhou Urban Planning Bureau.

from 1990 to 2003. Although the manufacturing sector is still the prime recipient of foreign direct investment, its share dropped significantly from 83 per cent in 1990 to 62 per cent in 2003. A large amount of capital has been invested in the service sector in the 1990s. The accumulated investment in the service sector between 1990 and 2003 was 11.89 billion USD, accounting for nearly 40 per cent of total foreign direct investment. Of the investment in the tertiary sector, the majority of the capital was invested in real estate, this pattern being particularly evident after Deng Xiaoping's visit to southern China in 1992. By the year 2003, over 63 per cent of FDI in the service sector in Guangzhou was found in real estate development (GZSB, 2004: 557). It is believed that urban transformation in China has been driven by real estate development and foreign investment has been one of the key actors driving the property-led urban engine (Ma, 2003). With the legitimization of land market in China, foreign investment has been actively involved in the upgrading of urban housing market and residential space. Moreover, the influx of foreign investment stimulates the demand for office space in better locations in the city center, which has accelerated the replacement of industrial land. As an essential external force, foreign investment has facilitated not only the process of industrialization but also the process of tertiarization in Guangzhou since its entry into China was allowed in 1979.

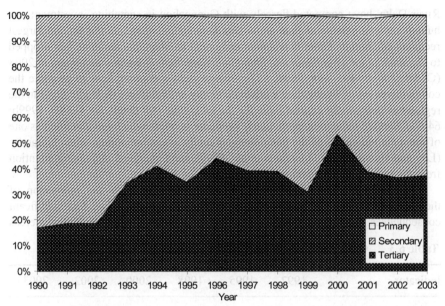

Figure 6.1   Foreign direct investment in Guangzhou Municipality, 1990–2003

Urban Developmental Strategies and the Growth of Services

To understand the growth of services and its consequence upon urban transformation in the Guangzhou metropolis, it is essential to analyze how local state (i.e. municipal government) has interacted with the new forces of marketization and globalization.

By virtue of Guangzhou and the Pearl River Delta's location, they are relatively remote, far away from the political center of Beijing. Such remoteness has important implications for the formation of state-market relation and local development of the region. Firstly, as the region is relatively remote from China's heartland but close to Hong Kong, the economic, social, and cultural impacts of Hong Kong on the region are significant. Secondly, the scalar reshuffling of power from the central to local government has given local governments and local people considerable flexibility in seeking development on the basis of their inherent comparative advantages. Local governments are able to respond to market demand and generate conditions conducive to attract foreign capital for local economic transformation and growth.

One of the new challenges faced by the municipal government of Guangzhou since the reforms has been the growing competition from nearby cities and towns in the Delta region. The share of GDP in the Delta generated by Guangzhou city has dramatically dropped from 34 to 23 per cent over the nine years from 1992 to 2000 (Lin, 2005: 293). The further expansion of market in China has placed Guangzhou in a more intensive competition circumstance as the policy advantages in the Delta have gradually diminished. Cities outside the Delta, particularly those in the Yangtze

River Delta, have competed fiercely with Guangzhou in snatching investment and human capital. Statistical data show that Guangzhou stood as the second largest recipient of foreign direct investment only next to Shanghai in 2000, but dropped to the sixth place after Shuzhou, Shanghai, Shenzhen, Qingdao, and Wuxi in 2003 (CSSB, 2001b: 495–498, 2004: 527–530). In such a context, how to promote the competitiveness of Guangzhou's urban economy and reinforce its leading role in the region has become the central issue for the local state (for detail, see GZSB, 1999b: 64-68). Service industries, particularly producer services which are identified as one of the main indicators of the positioning of a city in contemporary urban hierarchy (Ley and Hutton, 1991, Moulaert et al., 1997), have received increasing attention from the municipal government of Guangzhou.

The direct involvement of the municipal government in the promotion of the service sector as a means to meet regional competition has had significant consequences. As shown by Table 6.3, over the period from 1992 to 2003, whereas

Table 6.3   Guangzhou in the Pearl River Delta, 1992–2003

Indicator	1992	1995	1998	2000	2002	2003
Gross output value of industry (billion yuan)*						
Guangzhou	71.05	131.70	210.70	273.37	361.24	457.04
Delta	147.29	587.78	942.92	1286.62	1570.2	2109.98
GZ as % of Delta	48.24	22.41	22.35	21.25	23.01	21.66
Value added of tertiary industry (billion yuan)						
Guangzhou	23.38	58.94	92.63	124.95	167.13	188.31
Delta	–	162.77	255.87	329.29	426.56	493
GZ as % of Delta	–	36.21	36.20	37.95	39.19	38.20
Retail sales (billion yuan)						
Guangzhou	20.95	55.00	90.46	112.11	137.07	149.43
Delta	67.08	154.47	227.18	278.14	348.13	397.66
GZ as % of Delta	31.23	35.61	39.82	40.31	39.37	37.58
Value added of producer services# (billion yuan)						
Guangzhou	–	–	–	62.17	79.46	86.33
Guangdong	40.45	90.58	138.36	187.18	228.76	246.31
GZ as % of GD	–	–	–	33.21	34.73	35.05

Notes: * Gross output value of industry is measured at constant price of 1990;
For the reason of data availability, the proportion of Guangzhou in total value added of producer services of Guangdong province is calculated.
Sources: Guangdong Statistical Bureau (2001: 626); Guangdong Statistical Bureau (2004: 30, 567); Guangzhou Statistical Bureau (1999a: 231, 481, 612); Guangzhou Statistical Bureau (2004: 63, 453).

Guangzhou's share in total gross output value of industry of the Delta dropped significantly from 48.24 per cent to 21.66 per cent, its contribution to the Delta's value added of tertiary sector keep increasing steadily from 36.21 to 38.20 per cent, and its share of total retail sales of the Delta increased from 31.23 to 37.58 per cent. These findings suggest that the recent proportional decline of Guangzhou in GDP of the Delta has largely been in the manufacturing sector. While Guangzhou has lost its dominance in industrial production, it has been taking on new hub functions as the service and business center of the region. This contention has been further confirmed when the share of Guangzhou in total producer services of Guangdong province is interrogated. The proportion of Guangzhou in the province has been increasing steadily, and over 35 per cent of producer services were concentrated in the city by the year 2003. Service industries have become essential basic economic activities for the metropolis (Yan and Xu, 1999).

Conclusion

As a new and powerful driving force along with industrialization shaping the new urban economic landscape in 'post-socialist' cities (Lin, 2003), economic tertiarization has found its way to transform metropolitan development in China in the recent decades. With recent intrusion of globalization and marketization forces, the service sector has been growing dramatically in the Guangzhou metropolis over the last two decades and occupied a prime position in urban economy and land use arrangement. Consequently, Guangzhou has been undergoing a process of transformation from a city of industrial production to a service center in southern China. On its transformative pathway toward a service economy, the city government of Guangzhou has played a significant role in terms of state investment and its manipulation in urban planning.

An evaluation of the experience of Guangzhou suggests that the interaction of state and market has been fundamental to understanding the process of economic tertiarization and urban transformation. While the relationship between state-market interaction and the growth of services has been elaborated in voluminous literature, their applicability in the Chinese context has been problematic since the discourses and theses are derived from the cities embedded within a well-developed free market economic environment. As the experience of Guangzhou has shown, to explain the growth of service industries and urban transformation requires a comprehensive understanding of state-market relation which is historically and geographically contingent. Embedded within the Chinese state-market relation, the process of tertiarization has been distinct in at least two respects. First, the dramatic expansion of services in Chinese cities has not been a 'natural' process driven by free market forces, but subject to strong state manipulation for non-economic considerations. With a keen understanding of the fact that economic reforms or privatization might result in rising unemployment which could threaten social and political stability, the growth of services has been promoted by the Party-state as an outlet to accommodate the layoffs released from the SOE. Second, the unique central-local relation and the capability of the state in using market instruments in the Chinese context have significant implications for recent urban transformation. Strong localism and inter-urban competition have inevitably engendered rampant imitation and redundant construction, which are common phenomena in China and have led to a serious waste of resources. The empowered city government, however, has contributed to recent dramatic expansion of service industries in Guangzhou, which has helped to reposition the metropolis as a command and service center in the Pearl River Delta.

The findings of this study have raised several theoretical questions. Until recently, theoretical explanation for the growth of service industries and economic restructuring has not adequately addressed the role of state-market interactions. The case study of Guangzhou suggests that when embedded in various political economies, the ways in which economic resources are mobilized and allocated appear to be different, which results in the distinct patterns of urban development. Compared with markets, states are not generic, but rather, they vary dramatically in their internal structures and relations to society (Evans, 1995). Given that, the changing nature of the state and state strategy in the Chinese context needs to be conceptualized in further studies. The incorporation of state-market relation is believed to provide an important explanatory framework for understanding the growth of services in China.

References

Allen, J. (1988) 'Service industries: uneven development and uneven knowledge', *Area*, 20: 15–22.
Bell, D. (1973) *The Coming of Post-Industrial Society*, New York: Basic Books.
Block, F. (1994) 'The roles of the state in the economy', in Smelser, N.J. and Swedberg, R. (eds), *The Handbook of Economic Sociology*, Princeton: Princeton University Press, pp. 691–710.
Bryson, J.R. and Daniels, P.W. (eds) (1998), *Services Industries in the Global Economy (Volume I): Service Theories and Service Employment*, Massachusetts: Edward Elgar Publishing Limited.
Channon, D.F. (1978), *The Service Industries*, London: Macmillan.
China State Statistical Bureau (CSSB) (2001a), *China Statistical Yearbook (2001)*, Beijing: China Statistical Press.
China State Statistical Bureau (CSSB) (2001b), *China Urban Statistical Yearbook (2001)*, Beijing: China Statistical Press.
China State Statistical Bureau (CSSB) (2004), *China Statistical Yearbook (2004)*, Beijing: China Statistical Press.
Clark, C. (1940), The *Conditions of Economic Progress*, London, UK: Macmillan.
Coffey, W.J. (2000), 'The geographies of producer services', *Urban Geography*, 21 (2): 170-183.
Compilation Committee of China's Tertiary Industry Almanac (1993), *Almanac of*

China's Tertiary Industry (1993), Beijing: China State Statistical Press.
Compilation Committee of People's Republic of China Yearbook (2002), *People's Republic of China Yearbook (2002)*, Beijing: People's Republic of China Yearbook Press.
Cox, K. (1993), 'The local and the global in the new urban politics: a critical view', *Environment and Planning D: Society and Space*, 11: 433–448.
Daniels, P.W. (1982), *Service Industries: Growth and Location*, Cambridge, London, NY, New Rochelle, Melbourne, Sydney: Cambridge University Press.
Evans, P. (1995), *Embedded Autonomy: States and Industrial Transformation*, Princeton, New Jersey: Princeton University Press.
Fisher, A.G.B. (1939), 'Production, primary, secondary and tertiary', *Economic Record*, 15 (June): 24–38.
Friedmann, J. (1986), 'The world city hypothesis', *Development and Change*, 17: 69–83.
Friedmann, J. and Wolff, R. (1975), *The Urban Transition: Comparative Studies of Newly Industrializing Societies*, London: Edward Arnold.
Gong, H.M. (2002), 'Growth of tertiary sector in China's large cities', *Asian Geographer* 21 (1/2), 85–100.
Guangdong Statistical Bureau (GDSB) (1992), *Regional Economic Statistical Material of the Pearl River Delta (1980–1991)*, Beijing: China Statistical Press.
Guangdong Statistical Bureau (GDSB) (2001), *Guangdong Statistical Yearbook (2001)*, Beijing: China Statistical Press.
Guangdong Statistical Bureau (GDSB) (2004), *Guangdong Statistical Yearbook (2004)*, Beijing: China Statistical Press.
Guangzhou Statistical Bureau (GZSB) (1999a), *Guangzhou Fifty Years (1949-1999)*, Beijing: China Statistical Press.
Guangzhou Statistical Bureau (GZSB) (1999b), *Guangzhou Yearbook (1999)*, Beijing: China Statistical Press.
Guangzhou Statistical Bureau (GZSB) (2001), *Guangzhou Statistical Yearbook (2001)*, Beijing: China Statistical Press.
Guangzhou Statistical Bureau (GZSB) (2004), *Guangzhou Statistical Yearbook (2004)*, Beijing: China Statistical Press.
Hall, T. and Hubbard, P. (1998), *The Entrepreneurial City: Geographics of Politics, Regime and Representation*, London: Wiley.
Harvey, D. (1989), *The Condition of Postmodernity*, Oxford: Blackwell.
Hsu, R.C. (1985), 'Conceptions of the market in post-Mao China: an interpretive essay', *Modern China*, 11 (4): 436–460.
Hutton, T.A. (2004), 'Service industries, globalization, and urban restructuring within the Asia-Pacific: new development trajectories and planning responses', *Progress in Planning*, 61: 1–74.
Hutton, T.A. (2005), 'Service and urban development in the Asia-Pacific region', in Daniels, P.W., Ho, K.C. and Hutton, T.A. (eds), *Service Industries and Asia-Pacific Cities: New Development Trajectories*, London and New York: Routledge,

pp. 52–76.

Inman, R.P. (1985), *Managing the Service Economy: Prospects and Problems*. Cambridge: Cambridge University Press.

Kirby, M.W. (1986), *Deindustrialization and the UK Economy: Survey and Analysis*. Stirling: University of Stirling.

Ley, D.F. and Hutton, T.A. (1991), 'The service sector and metropolitan development in Canada', in Daniels, P.W. (ed.), *Services and Metropolitan Development: International Perspectives*, London and New York: Routledge, pp. 173–203.

Lin, G.C.S. (1999), 'State policy and spatial restructuring in post-reform China, 1978-95', *International Journal of Urban and Regional Research*, 23: 670–696.

Lin, G.C.S. (2002), 'The growth and structural change of Chinese cities: a contextual and geographic analysis', *Cities*, 19 (5): 29–316.

Lin, G.C.S. (2003), 'Toward a post-socialist city? Economic tertiarization and urban reformation in Guangzhou metropolis, China', *Eurasian Geography and Economics*, 44 (8): 507–533.

Lin, G.C.S. (2004) The Chinese globalizing cities: national centers of globalization and urban transformation. *Progress and Planning*, 61(3): 143–157. London: Elsevier.

Lin, G.C.S. (2005), 'Service industries and transformation of city-regions in globalizing China: new testing ground for theoretical reconstruction', in Daniels, P.W., Ho, K.C. and Hutton, T.A. (eds), *Service Industries and Asia-Pacific Cities: New Development Trajectories*, London and New York: Routledge, pp. 283–300.

Lo, C.P. (1994), 'Economic reforms and socialist city structure: a case study of Guangzhou, China', *Urban Geography*, 15 (2): 128–149.

Logan, J. and Molotch, H. (1987), *Urban Fortunes: the Political Economy of Place*, Berkeley, CA: University of California Press.

Ma, L.J.C. (2002), 'Urban transformation in China, 1949–2000: a review and research agenda', *Environment and Planning A*, 34 (9): 1545–1569.

Ma, L.J.C. (2003), 'Some reflection on China's urbanization and urban spatial restructuring', Urban China Research Network Workshop on 'Urban Studies and Demography in China', MN: Minneapolis. April 30, 2003.

Marshall, J., Damesiek, P. and Wood, P. (1987), 'Understanding the location and role of producer services in the United Kingdom', *Environment and Planning A*, 19: 575–595.

Marshall, J.N. and Wood, P.A. (1995), *Service and Space: Key Aspects of Urban and Regional Development*, New York, Singapore: Longman.

Moulaert, F., Tödtling, F. and Schamp, E. (1995), 'The role of transnational corporations', *Progress in Planning*, 43 (2/3): 107–121.

Moulaert, F., Scott, A.J. and Farcy H. (1997), 'Producer services and the formation of urban space', in Moulaert, F. and Scott, A.J. (eds), *Cities, Enterprises and Society: on the Eve of the 21st Century*, London and Washington: Pinter.

Oi, J.C. (1992), 'Fiscal reform and the economic foundations of local state corporatism in China', *World Politics*, 45 (1): 99–126.

Oi, J.C. (1995), 'The role of the local state in China's transitional economy', *The*

China Quarterly, 144: 1132–1149.

O'Neill, P.M. (1997), 'Bringing the qualitative state into economic geography', in Lee, R. and Wills, J. (eds), *Geographies of Economies*, London: Arnold, pp. 290–301.

Putterman, L. and Dong, X. (2000), 'China's state-owned enterprises: their role, job creation, and efficiency in long-term perspective', *Modern China*, 26 (4), 403–447.

Sassen, S. (2000), *Cities in a World Economy (Second Edition)*, Thousand Oaks, London, New Delhi: Pine Forge Press.

Sassen, S. (2001), *The Global City*, NY, London, Tokyo: Princeton University Press.

Scott, A.J. (ed.) (2001), *Global City-Regions: Trends, Theory, Policy*, Oxford: Oxford University Press.

Sit, V.F.S. and Yang, C. (1997), 'Foreign-investment-induced exo-urbanization in the Pearl River Delta, China', *Urban Studies*, 34 (4): 647–677.

Smart, A. (1997), 'Oriental despotism and sugar-coated bullets: representations of the market in China', in Carrier, J.G. (ed.), *Meanings of the Market: the Free Market in Western Culture*, Oxford: Berg, pp. 159-194.

Solinger, D.J. (1983), 'Marxism and the market in socialist China: the reforms of 1979-1980 in Context', in Nee, N. and Mozingo, D. (eds), *State and Society in Contemporary China*, Ithaca and London: Cornell University Press, pp. 194–222.

Solinger, D.J. (1993), *China's Transition from Socialism: State Legacies and Market Reforms, 1980-1990*, Armonk and London: M.E. Sharpe.

Tickell, A. (1999), 'The geographies of services: new wine in old bottles', *Progress in Human Geography*, 23 (4): 633-639.

Tickell, A. (2001), 'Progress in the geography of services II: services, the state and the rearticulation of capitalism', *Progress in Human Geography*, 25 (2): 283–292.

Wu, F. (2002), 'China's changing urban governance in the transition towards a more market-oriented economy', *Urban Studies*, 39 (7): 1071-1093.

Xia, M. (2000), *The Dual Developmental State: Development Strategy and Institutional Arrangements for China's Transition*, Aldershot, Brookfield: Ashgate.

Xu, J. and Ng, M.K. (1998), 'Socialist urban planning in transition: the case of Guangzhou, China', *Third World Planning Review*, 20 (1): 35–51.

Yan, X. and Xu, X. (1999), 'Guangzhou chengshi jiben-feijiben jingji huodong de bianhua fenxi (Changes of the basic-nonbasic economic activities in Guangzhou: a re-consideration of the economic base theory of urban development)', *Dili Xuebao (Acta Geographica Sinica)*, 54 (4): 299–308.

Yang, F.F. (2004), 'Services and metropolitan development in China: the case of Guangzhou', *Progress in Planning*, 61 (3): 181–209.

Zhang, W. (2000), *Transforming China: Economic Reform and its Political*

Chapter 7

SIA and Singapore: Competition, Changes in Organization and Technologies and the Impacts on Economic Development

Shuang Yann Wong[1]

Introduction

Singapore, with a GNP per capita of about US$22,000 is the richest developed country in Southeast Asia despite the lack of natural resources. Its success is often explained in terms of the developmental state model, but the power of the State is now increasingly challenged by external forces, albeit in the past it has intervened effectively in transforming the formerly *entrepot* economy into a global centre of production and distribution of high value-added goods and services. The commanding heights of transnational corporations (TNCs), the increased accessibility and mobility of information and communication technologies (ICT), capital and labour have interrupted the development of many developing countries. The liberalization measures emphasized by international agencies further complicate the agenda of change as competition intensifies. Having moved from the Third World to the First World, can Singapore continue to rely on the state developmental approach to sustain its competitiveness in the new global economy?

To meet the challenges of the new millennium, various incentives and policies have been enforced recently. As selected manufacturing clusters are promoted, increasing emphasis is now placed on the services sector. In banking and finance, liberalization and regulatory measures have been implemented to deepen the integration of Singapore into the global economy (Wong, 2005, a and b). Transport and communication and tourism are important revenue-earners in the services sector. The Singapore International Airlines (SIA) carries the branded image of the city-state's transport services sector, supplying the key air transport services while complementing the needs of the closely related international tourism sector. Its role will be enhanced further as the Republic sets out to become a global business and aviation hub. Constrained by the lack of space and a small domestic market, SIA

---

1   National Institute of Education, Nanyang Technological University, Singapore.

may not have achieved today's prominence without the State's support. However, increasingly, forces that are affecting the airline's performance are now beyond the power of the State. The 1997 regional crisis, the outbreak of SARs, international terrorism, pandemic avian flu and the intermittent rise in oil prices have slowed the progress of the national economy. The consequent drop in passenger flows and profit margins forced SIA to take the unprecedented move in 2003–2004 of retrenchment which antagonized its labor force.

Taking the experience of SIA as an example, this chapter revisits the relevance of the developmental state model for explaining the current problems, issues and constraints related to the airline and their impact on the future development of the Singapore economy. Restructuring and integrating the domestic economy into the global economy require more than what a State can do. The sustenance of SIA as a global player will require the cooperative support of not just the State but also other related actors in the industry. This chapter argues that, in the context of Singapore's changing socio-economic structure, the State-centric developmental state strategy may not lose its currency. A new formula of development based on a cooperative relationship of the State with the stakeholders or compatible partnership will navigate the course of development of the national economy. This chapter tries to explore this relationship by observing the existing links between SIA as an organization, the State and the airline's employees. The analysis is organized as follows. The first section reviews the Singapore economy and the directions of change and the air travel industry. A conceptual framework is then used to highlight the significance of the cooperative relationship between the State, industry organization and labour force in sustaining the competitiveness of a global service provider like SIA. Third, the development process and strategies of SIA in branding its products and internationalizing its markets are presented. Next, the working relations of SIA with the State and labour are examined followed by the discussion of the problems of SIA today, the strategic responses and the relevance of the developmental state model. The airline's future prospects, suggested research and concluding remarks are provided in the last section.

The Singapore Economy: Directions of Change

Since its independence in 1965, Singapore has weaned itself from the dependence on the Malay Peninsular hinterland. Making itself the most open economy in the region, it planned successively to diversify into the manufacturing and modern services sectors. With an area of 655 square kilometers and no significant natural resources other than a deep water harbor and good geographical location, the government launched the aggressive foreign direct investment (FDI)-led, export-oriented industrialization programme. To ensure the success of the programme, the State exerts its power to oversee the development of efficient physical and social infrastructure facilities which include the construction of internationally linked maritime port and airport, industrial parks, public utilities, public housing and schools.

To maintain political stability and harmonious industrial relations for the attraction of FDI, the State forms the tripartite National Trade Union Congress (NTUC) as a means to discipline the work force and to safeguard workers' welfare. Most workers are NTUC members and labor disputes are uncommon. To accommodate the need of professional skills and technical expertise, the State adopts a liberal immigration policy for the employment of foreign professionals, explaining the relatively large number of SIA's expatriates staff.

The effectiveness of these policies and incentives is shown by the continuous inflows of FDI, often the largest in the region until the recent rise of China. In the last three decades, Singapore's economic growth is fueled largely by TNC FDI that engaged in high-tech and capital-intensive manufacturing industries. These manufacturing TNCs undertake a large part of the services needed to support their global value chains, especially in trade and finance (Daniels, 2005). The State dominates in key services such as shipping, aviation, finance and telecommunication, sometimes in direct competition with the private sector such as in finance and banking (Wong, 2005b). The State-supported transport and communications sector has enabled Singapore to become a major trans-shipment hub, a global warehousing and distribution centre and a significant location for TNC Regional/Operational Headquarters in the Asia Pacific region (Ho, 2000, Yeung et al. 2001). Transport and communication is not just an important service export sector, it plays a vital

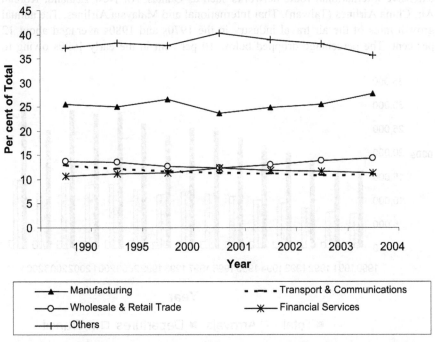

Figure 7.1 **Economic structure of Singapore by industry distribution of gross value product, 1990–2004**

role in complementing or servicing the needs of other exports too, namely trade and commerce, manufacturing, finance and banking and international tourism (Daniels, 2000). Its malfunction can mar the overall performance of the city-state that is externally dependent on foreign markets and the mobility of capital, labor, management know-how and technology for the exports of its goods and services. Transport and other services have helped diversify the country's economic structure as the share of the manufacturing sector in its GDP declined to about a quarter while that of the services increased to about two-thirds. Transport and communications consistently contributes about 10–12 per cent to the GDP (Figure 7.1) and an average of 4–5 per cent to total goods and services exports for the period 1990–2004. Air transport in particular plays an active role as it accounts for about 60 per cent of international travel. Between 1990 and 2004, air passenger traffic doubled from about 15 million to 30 million, though with a shortfall of about 5 million in 2003. Generally, the number of arrivals and departures is almost the same (Figure 7.2). As competition in the new global economy intensifies, can SIA as the key air transport player of the city-state sustain its position both as a leading service provider and export service?

Some of Asia's top airlines, most of which are national carriers, have matured into major world players during the booming 1970s. SIA, Japan Airlines (JAL) and Cathay Pacific are examples. Some achieved enormous profits on the basis of extensive international route networks such as Qantas, Air New Zealand, Korean Air, China Airlines (Taiwan), Thai International and Malaysia Airlines. The annual growth rates of the air travel industry in the 1970s and 1980s averaged about 12 per cent. The percentage dropped below 10 per cent in the early 1990s owing to

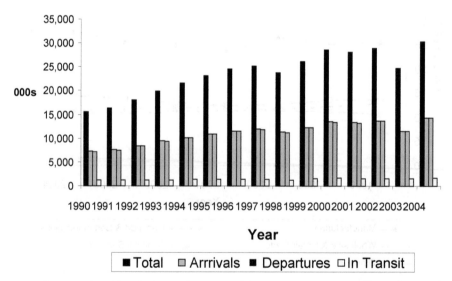

Figure 7.2   Passenger arrivals, departures and in transit by air, Singapore, 1990–2004

the 1991 Gulf War and the subsequent recession, still high compared to the US and international growth rates. Recovery was stunted by Japan's sluggish growth as it accounted for nearly 60 per cent of the international air traffic in Asia. Before full recovery, the region was hit by the 1997 currency crisis, thereafter the bird flu and the SARs. Many Asian airlines suffered losses. For example, during the period 1993–94, JAL lost about US$234 million while Cathay Pacific and SIA saw their margins reduced from 20 per cent to 12 per cent (Labich, 1994). The estimate of the International Air Transport Association (IATA) on the region's passenger growth was too optimistic when it estimated that international scheduled passengers would grow by an average annual growth of 7.1 per cent to reach the number of 522 million, and that Asia would capture about half of the worlds' international air travel by 2010. Such positive prospects must be set against the unequal distribution and competition within the international air travel market. Australia's recent rejection of SIA's request to participate in the lucrative Sydney/Melbourne-Los Angeles/San Francisco route monopolized by Qantas and United is a good example. At the same time, deregulation in the USA has intensified competition, compelling airlines to undertake cost-cutting and price slashing in the early 1990s, especially on some of the more popular and hotly contested routes. In Asia, towards the late 1990s, to sustain competitiveness, JAL and Cathay Pacific reduced their workforce, hired more expatriates and based new staff in their respective home countries to reduce costs.

Over the years SIA has grown into a premium quality global service provider through the substantial support of the State. The city-state's unique developmental state strategy has institutionalized the regulatory rules and governance regime that cultivate and sustain SIA's branded prestige. Like other airlines, SIA made the necessary structural adjustments under the pressures of globalization, liberalization and competition. It became partly privatized giving the impression of the declining role of the State. In reality the State's role remains strong not only in maintenance of infrastructure and route expansion but also in industry/labor relations as the following analysis unfolds. This chapter argues that it is not the role of the State *per se* that defeats the validity of the developmental state strategy. In the context of sustaining the competitiveness of a national carrier as a profitable branded global service enterprise, it is crucial to maintain a cooperative relationship or compatible partnership among the actors, *viz.* the State and the service providers which comprise the industry organization and labor force. The latter in particular play the influential role as they are in direct contact with the consumers; their failure (intentionally or unintentionally) to provide the services satisfactorily will affect the market and finally the contribution to the national economy.

Services and Economic Development: A Conceptual Framework

In economic geography, Christaller's central place model has long been used to explain the spatial hierarchy of service provision. While market dynamics and scale economies remain important to the neo-liberal school of thought for explaining the spatial pattern of service provision today, the social and cultural perspectives have gained credence. The ethnographic studies of market differentiation are examples. Considering Singapore's geographical and resource constraints, its open and outward-oriented economy, and its state-centered policy making, cultural reductionism or developmentalism is too narrow to succinctly explain the processes that have moulded SIA's brand and the strategy that is used to effect the organizational and institutional adjustments to sustain its position as a branded top global service provider. This paper takes a spatio-temporal perspective that incorporates an actor-institutional relationship to explain the unique case of SIA, in particular how it has evolved from a national carrier of a city-state to a global service enterprise today through scaling new heights in the past decades.

In the provision of a service, several actors can be envisaged – consumers, service providers (organizations, management and employees), politicians and policy-makers (Figure 7.3). The effective delivery of a service would require the cooperation of these actors which in turn is subject to the power relations and complementarities between them. The State has the monopoly of sovereign right that legitimates the use of power by politicians to regulate, legislate, tax, define and enforce the rules of the game. The politicians control this power and discharge the fundamental responsibilities of the State. Policy-makers also exercise the power of the State. In some countries, politicians are also policy-makers. Politicians generally chart the directions, but policy-makers set the fundamental rules of the game within which service providers operate – by regulating entry, enforcing standards, and determining the conditions under which providers receive public funds. Service providers can be public or private organizations, professionals or non-professionals.

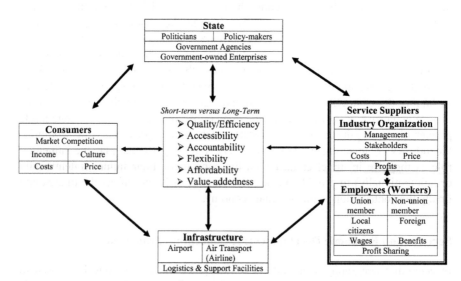

Figure 7.3   A conceptual model of sustainable service provision, major actors and key relationships

Service provider organizations can be state- or privately-owned or autonomous public enterprises as in the case of SIA. The head of the provider organization makes internal policies specific to the organization and is directly responsible for management. Complications may set in when the head is also the policymaker as it is difficult to have a clear delineation of policy-making and direct production responsibilities. Direct service providers are those who come in direct contact with consumers. In the airline industry they include the pilots, cabin crew, engineers, food and beverage caterers and others. The consumers form the base of the market. Competition intensifies as providers inject new products into the services or cut costs to influence prices so as to attract the increasingly mobile customers who have become more sophisticated in values, beliefs, identities and capabilities.

The marketability of a service and its impact on a national economy is therefore dependent on the relationship between service provider organization, the frontline service workers, politicians, policy-makers and consumers. This relationship can be a social contract. The State can provide resources and delegates powers and responsibility to politicians and policy makers for collective objectives. Rewards and penalties may be ensured depending on performance. The service organization's relationship with frontline workers touches on management issues that include recruitment, training, wage levels, welfare benefits, performance bonuses and others. When the service is market-dependent, the relationship of service providers and consumers is vital as the latter can exercise influence over their performance. To ensure standards and performance, service providers rely on the efficiency and professionalism of their frontline workers. The failure of delivering the services directly can result in discontinuation of the service as demand ceases. As a privatized enterprise, an airline has the autonomy to manage its services but often requires the support of the State in many fundamental areas to sustain the service quality, market share and stability. These include negotiation on sovereignty of flyover air space, landing rights, injection of funds for construction, acquisition and maintenance of airport facilities and aircraft, security enforcement, immigration controls and policies on staff employment. Additionally, a strong State can provide a clear vision and mission for service provision, not one that is merely process-oriented, internally incoherent and externally anachronistic to the needs of the time.

Airline services are both discretionary and transaction-intensive. Discretionary measures are enforced by administrative and bureaucratic controls to monitor the quality and efficiency of logistical tasks. Transaction-intensive strategies are to expand networks, trap traffic volume and upgrade customer services to satisfy demands from different market segments. These measures may help to avoid the characteristics of failure cases such as inaccessibility, dysfunctionality, low technical quality, weak market responsiveness, poor training and stagnant productivity. Affordability and accountability are also important especially if the airline is also a national flag carrier and has thick institutional links with the State. An effective trilateral communication between State, airline industry (management) and labor and a shared vision are essential to maximize desirable outcomes. The appointment of government officers or bureaucrats to sit in the board of directors or executive council of the airline may help cement the contract between the State and the airline, but not the contract between the airline and labor unless representatives from labor also sit in the airline's executive committee.

A strong contract between the organization and labor is essential. The management of an organization, in using its autonomy and discretion, may sometimes overlook the needs and interests of staff that have vital links with customers. The relationship of the management with the staff is not confined to short-term training and routinized supply of basic social amenities. The relationship thrives on the long-term social contract characterized by trust, mutual respect and accountability and is not simply measurable in terms of wage levels and employment benefits. A breakdown in the social contract can occur when the management is too frequently seeking improvements in technical efficiency, market expansion, costs-cutting and profit attainment, and it may undermine performance to the detriment of both parties. State intervention may become necessary to rectify communication breakdown and to clear any misunderstandings that may jeopardize the interests of the State. The existing institutional arrangements may have to be changed to strengthen the weak link before staff demand for reforms that may sap the competitiveness of the service provider. Instead of using short-term solutions, incremental activities through pragmatic improvisation may be more effective in creating the conditions for reforms in the long run.

How sustainable is SIA's competitiveness and its hub position in the face of regional and international competition? What changes in the development strategy and management system have affected the relationships between the management, the State and the key direct service providers? How does the State lend its hand or raise the baton in pre-empting any breakdown in the relationship? Before examining the recent events that seem to destabilize the future of the airline and challenge the role of the State, it is useful to look at the background leading up to the national carrier's strength today.

SIA: Process of Change, Constraints, Problems, Competition and Strategy

The roots of SIA date back to 1947 when it first started scheduled services between Singapore and the key cities of Malaya as *Malayan Airways*. By 1955, it offered international services between major cities in Southeast Asia. In 1963, it was renamed *Malaysian Airways* to signify the formation of Malaysia comprising Malaya, Singapore, Sabah and Sarawak. Although Singapore became independent and withdrew from Malaysia in 1966, the airline was jointly controlled by the governments of Malaysia and Singapore. In 1967, the airline was renamed *Malaysia Singapore Airlines* (MSA). Given its limited territorial expanse and resource constraints, Singapore specialized in international flights while Malaysia concentrated on domestic and regional flights. As differences in management and operations of the airline increased, Malaysia established its own airline in 1971. In 1972 Singapore started its own airline, *Mercury Singapore Airlines (MSA)* which was later revised to *Singapore International Airlines* (SIA). Having retained a fleet of the B707s and B737s, SIA continued to serve the entire international network formerly served by MSA and also actively refurbished it. It was the first in Southeast Asia to order a Boeing 747 and in 1973 set up a subsidiary, Singapore Airport Terminal Services (SATS), to perform ground handling, catering and related tasks.

As the business of SIA expanded, the old Paya Lebar airport near the CBD could not cope with the demand. Changi International Airport located in the northeastern part of the island-state, was opened in 1981. At the time it was the largest in the region and could handle 10 million passengers per year. Changi is also equipped with logistics supporting, maintenance, repairing and re-fueling services, warehousing, in-flight catering kitchens and the largest computer installation in Singapore. These airport facilities are meant to capture extra income by providing services to foreign airlines which often contract with local airports for local maintenance capacity to save costs. Being a small city-state without any domestic routes to monopolize, SIA concentrates on resource enhancement, route expansion and marketing its services internationally. To supply a premium service and a uniquely different image, SIA created the Singapore Girl as its icon to evoke the very best in Oriental charm and hospitality as the representation of the airline's quality and friendly service. The culture of service excellence through rigorous training and service-first policy forms the bedrock of the organization. By combining routes inherited from previous joint venture with its own, SIA had links with 50 cities worldwide including the capital cities of East and South Asia, and major cities in Australia, New Zealand, Western Europe and North America. By 1980, SIA was offering nine weekly flights across the Pacific between major hubs in Asia (Tokyo, Hong Kong, and Taipei), and three main cities in the USA (Los Angeles, San Francisco and New York).

By 1982, SIA had generated 18,081 million Revenue Passenger Kilometers (RPKs), moving ahead of competitors such as Qantas and Swissair. Between 1973 and 1983, total revenue grew six-fold from S$389 million to S$2,621 million, with most sales coming from the developed West. Its passenger load factors were consistently above 70 per cent for the greater part of the period and staff strength more than doubled from 4,906 to 10,655. In 1988, it was ranked the world's most profitable airline by *International Business Week*. The entry of some USA airlines into the region compelled other Asian carriers to undertake cost-cutting measures and neglect service quality. SIA, on the contrary, focused on cost-saving and upgrading at the business unit level such as in aircraft maintenance, service improvement, catering and security. It also relocated its data processing functions to low-cost locations in India and China. These measures did result in improved cost savings and greater profitability for SIA for the period from 1970 to 1989. To capture the long-haul business market and regional tourism, SIA went into aggressive capacity building. In 1978, it acquired 13 B747s and six B727s, then the largest commercial airplane order in the world. In 1989, SIA together with Delta Airlines and Swissair formed the Global Excellence Alliance that serves 300 cities in more than 80 countries, enabling SIA to reach more destinations in the USA and Europe.

The year 1990 was battered by recession, oil price hikes, high interest rates and the Iraq-Kuwait war which sharply reduced passenger loads on most international routes. Many airlines suffered heavy financial losses and started retrenching, but SIA increased its advertising budget and reported the highest operating profit of any airline in the world in 1990. For the period of 1990–1991, its revenues and load factor declined marginally from S$5.09 billion to S$4.95 billion, and from 78.3 per cent to 75.1 per cent respectively, but the number of passengers rose to 8.1 million partly because of the jump in fleet size. In 1991, SIA maintained a fleet of 29 747s and 14 Airbus-310s, and was among the ten largest airlines in the world in terms of international ton-kilometers of load carried. In the same year, it became a member of IATA, gaining a voice in international forums, and greater access to the organization's technical expertise and accredited sales agents. More upgrading was done at the terminals. At Terminal 2, automatic verification of check-in and check-out details of passengers and a sky-train that ferries passengers and baggage between terminals were introduced. In 1991, the carrying capacity of Terminal 1 was increased to 44 million passengers a year. Integrated quality services were provided at the airport to visitors in transit including a gymnasium, a conference hall, an exhibition hall, two business centers, 4 full-service banks, 20 restaurants, more than 100 shops and many others. In-flight technological advances were also made such as installing a small TV screen for each First and Business Class seat. To provide travel agents with extended services for airline and hotel reservations, ground arrangements and regional travel news, SIA formed the Abacus[2] with others. More than 100 carriers, 80 hotel chains and others are using the Abacus.

The globalization of the services has resulted in a network of subsidiaries and affiliated firms which separately run and manage their respective strategic units. Its subsidiary at Changi, Singapore Airport Terminal Services (SATS) handles passengers, baggage, cargo, and mail at all 63 airports in the SIA network. Its affiliate, SATS Catering, has a S$170 million kitchen complex that can produce 30,000 meals daily. The Service Quality Centre takes care of crew training, the Singapore Engineering Company (SIAEC) supplies aircraft maintenance and upgrading services. Other subsidiaries include Singapore Aircraft Leasing Enterprise (SALE) and Singapore Flying College. Airport activities accounted for about 10–15 per cent of SIA's total revenues. To capture more regional markets, a subsidiary called TradeWinds was formed which was later renamed SilkAir. SIA's air cargo services also expanded, forming about 20 per cent of its revenues. Between 1991 and 1996, air cargo at Changi grew at an average annual rate exceeding 15 per cent and about 62 per cent of it was trans-shipment cargo. SIA was the fifth largest after Lufthansa, Air France, Korea Airlines and JAL in terms of cargo carried in the world. The Changi Airport and SIA airline services are mutually reinforcing as they attempt to carve a niche in the region's air transport sector. In 1996, SIA posted group profits of S$732 million.

---

2   These include Cathay Pacific, China Airlines, DragonAir, Malaysian Airline, Philippine Airlines, Royal Brunei Airlines, SIA's SIlkAir and WorldSpan Global Travel Information Services.

In 1997, many accolades were won including Top Airline by readers of the *Asian Wall Street Journal*, world's best airline for the fourth successive year in the Zagat airline survey and Executive Travel's coveted Airline for the Year 1997. Despite the 1997 regional currency crisis, SIA made profits in 1998.

The second millennium ushered in a period of global terrorism after the September 11 attack in 2001. Most airlines struggled with the increased costs of added security and higher insurance premiums. In East Asia the problems were compounded by the SARs outbreak. Ironically, the ease of air travel spread the disease to over 20 countries. The airlines in East Asia lost money for six months as the fear of SARs stopped travel to affected countries, especially China, Hong Kong, Vietnam, Taiwan and Singapore. Other problems included competition from regional counterparts, the entry of low-cost carriers, the introduction of new aircraft and rising fuel prices. The most difficult of all are probably the pressing demands by staff for higher wages and continuation of welfare benefits. Globally, in 2003, the air transport industry worldwide shed about 100,000 jobs. The IATA estimated that airlines lost about US$30–40 billion over the period 2001–2003 as a result of competition, global terrorism and SARs. SARs-hit Asia saw a 44.8 per cent drop in air passenger traffic in Jan 2003 against a global backdrop of an 18.5 per cent slump (*Straits Times*, 7 Feb, 2003). In 2003, SIA lost S$204 million in April alone or more than S$6 million a day. As former CEO of SIA, Cheong Choong Kong stated in his farewell speech:

> These are terrible times. I wish my retirement did not have to coincide with the most difficult period in SIA's history, for never before have we been so battered by forces beyond our control or faced wage cuts and retrenchments on the scale envisaged. (*Straits Times*, 23 May, 2003).

Additionally, Singapore's hub position and SIA may be threatened by the arrival of new aircraft, the expansion and emergence of regional and new airlines and airports. The new Airbus A340–500 and the 550-passenger A380 allow airlines to bypass Singapore. For example, airlines flying from Europe *en route* to Australia can now choose to stop over at further points, like Dubai instead of Changi. Meanwhile competition from Thai Airways and the Emirates increases. The equally strategically located and oil-rich Dubai is going to charge 10 per cent less whatever Changi charges. New airports at Hong Kong, Seoul, Beijing, Shanghai and Guangzhou may not compete directly with Singapore in the short run, but direct competition from Bangkok's new airport cannot be ignored as Bangkok is better positioned geographically to serve both the Europe-Asia and the Europe-Australia/New Zealand route. Changi's current advantage over Bangkok is due to SIA's sharing of the Kangaroo route with BA and Qantas which hub in Changi. China will be the second largest aviation market after the USA in the next 20 years and potentially competes with Singapore and Changi's regional hub services. These threats can be real considering the ageing of the city-state's existing tourism products, its lack of natural attractions and the dependence on FDI and other foreign inputs such as labour and energy fuel.

Besides contending with other major airlines, SIA also faces competition from low-cost carriers (LCCs) which are taking business away from full-service carriers. Air transport is increasingly being viewed as a commodity, the cheaper it is, the more likely it is to attract customers. Relying on full-service airline's revenue to sustain growth is not sufficient as the LCCs are commoditizing the tourist class at a level much lower than that of full-service airlines. In lowering the price to the level of the LCCs, the profit margins of full-service carriers such as the SIA may dip. LCCs have currently captured 25 per cent of the market share in the USA, 10 per cent in Europe, and 40 per cent in Australia, but only 5 per cent of Asia's intra-regional market (*Straits Times*, 5 May, 2005). The skies of Asia are much less free than in the US and Europe. The Asian LCCs depend on bilateral agreements to meet strict regulatory requirements. Singapore and Thailand have a free skies deal which allows Singapore LCCs fly to Thai destinations but access to other routes in Asia remains difficult.

To sustain competitiveness, SIA has started cost minimization and international-izing its operations through outsourcing and diversification since the 1980s and hedges 30–50 per cent of its fuel needs. To counteract the 2003–2004 downturns caused by the SARs, SIA reduced the number and frequencies of its flights by almost 40 per cent, especially to SARs-affected areas (*Straits Times*, April 30, 2003). The fleet size was contracted by sending Boeing 747–400s and Airbus for retrofitting and routine maintenance, leaving less than 80 Boeing aircraft in active services. Ten units of A380 have been purchased recently to promote long-haul non-stop flights to key points in North America such as Los Angeles and New York as well as bringing in passengers. To enhance Changi's attractiveness, about S$2 billion will be spent on upgrading the three terminals to create ambience and more space for indoor recreation, retailing and food and beverage outlets which currently account for about one-third of the airport's annual revenue. High airport standards will be maintained on immigration clearance, ground handling, connecting flights as well as security and crisis management such as SARS and terrorism. Notwithstanding the long list of services, Changi has the second lowest airport charges in Asia and offers various incentives to encourage airlines to make use of its services.

With the state's support, undoubtedly SIA has succeeded in improving its hardware, namely the airport, traffic volume, fleet and route expansion, quality maintenance and other infrastructure upgrading. Its software management in terms of labor relations is less desirable. The aviation industry provides about 220,000 jobs. The airport employs about 35,000 permanent workers and 55,000 contract or temporary workers. SIA's staff numbers about 12,000; of which about 1,700 are pilots and 6,400 are cabin crew. The social and economic implications are great if SIA declined into a mediocre carrier and Changi lost its role as a major hub. In early 2004, about 100,000 were unemployed; the stretched social fabric would tear if more were retrenched by SIA and Changi (*Straits Times,* Jan, 10, 2004). SIA was also caught unprepared when its frontline staff displayed their grievances in 2003–2004. The lack of flexibility in managing labor, not so much in skills training but in safeguarding staff welfare benefits during period of crisis and long-term interests, has unveiled the fragile social contract of SIA with its staff, the patently weak corporate culture despite it being a global service enterprise.

SIA and the Role of the State

The failure of SIA in nipping the 2003–2004 union problems in the bud may be explained partly by its dependence on the State for support and labor control. Throughout SIA's history, its goals have been oriented towards the development stage of the country and the directions set by the State. Before 1981 the objectives of the airline reflected corporate business purposes and the promotion of the growth of the domestic economy. After 1981 commercial priorities and international expansion were emphasized. The significant role that SIA played in supporting the nation's tourism, trade and commerce and business sectors exemplifies the cooperative and complementary relationships with the State and the key institutions. However, the State taxes SIA like any other commercial organization, and it has to use its own profits to invest in new equipment without incurring significant debt. To finance fleet acquisitions, management tapped internally generated funds and issued shares. The government may provide guarantees on loans for expensive aircraft purchases. The State oversees the management-labor relations and mediates labor disputes to pre-empt any loss of worker days.

The State actively engaged in SIA's initial development such as building the Changi Airport and providing the associated logistics and capital investment. The State helps in immigration control, negotiation of landing rights, international routes and frequency of flights. The State succeeded in signing an Open Skies agreement with the USA and in negotiating new routes to provincial capitals in China. Though a majority shareholder, the State generally does not interfere in the decisions of SIA. However, it monitors decisions concerning shareholder matters and periodically reviews its performance and future plans. It may approve or provide equity capital or shareholder loans for the development of projects beneficial to the country. To finance its operations, SIA primarily made use of bank loans. Its initial lease financing was arranged by Morgan Grenfell and DBS was the main financier. After privatization, SIA was listed in the Singapore Stock Exchange in 1985 with a floatation of 100 million shares and DBS was the managing underwriter. SIA manages its own capital funding and debt/equity ratio. Although there are many foreign and local banks in Singapore, DBS arranged much of SIA's financing as in the case of other GLCs in Singapore. The State generally does not guarantee any debt, but is concerned with SIA's level of risk.

The primary means of State control in the management of SIA was via the appointment of civil servants to the airline's Board of Directors. The State's *ad hoc* intervention is to support and complement company operations. The administration of SIA, led by civil servants, operates on the basis of a cooperative network of the Republic's private and public institutions. This interface with the government is an advantage as the civil servants form an elite corps of professionals that is part of the team in constant communication with the national leadership to discuss the development policies and directions of the country. For example, Mr. Pillay was once the SIA Chairman as well as the managing director of the Government Investment Company and the Monetary Authority of Singapore, the Chairman of government-linked Development Bank of Singapore (DBS) and Temasek Holdings, among other positions. There are also interlocking directorates uniting SIA and the Civil Aviation Authority of Singapore (CAAS). In 1984, SIA's Deputy Chairman was also STB's Deputy Chairman. The arrangement is to dovetail the goals of SIA with those of the State.

SIA and Labour Relations: Souring Ties, Flexibility and the Gaining of Trust

With rising living standards in Singapore, SIA finds it difficult to attract the young and the best staff and grapples with soaring costs without jeopardizing the quality of its service. Having suffered its first-ever quarterly loss in the second quarter of 2003 as a result of the SARs, avian flu and global terrorism, and following other major airlines in the West, SIA cut salaries, imposed compulsory unpaid leave on pilots and cabin crew, and slashed jobs by up to 20 per cent. These measures can achieve savings of about S$30 million which were insignificant compared to the huge wage bills of S$14 billion for the year ended in March 2003 arising from the 3.23 months of staff profit sharing bonus (*Straits Times*, May 31, 2003). SIA management intended to slash staff-related costs by more than S$200 million of some 14 per cent to shrink the wage bill. The pilots form about 14 per cent of the airline's total staff, but take up almost a third of the total wage bill. The corresponding percentages for the cabin crew are 47 and 34 respectively. SIA intended to cut pilots' wages up to 22.5 per cent. Among the four SIA unions, the pilot union had a more turbulent relationship with SIA. Owing to the high costs of training a pilot ($500,000–$700,000), airlines including SIA tend to poach a pilot from a rival rather than train one, making pilots more mobile and giving them the bargaining edge. Unhappy over the pay-cut and perceived preferential treatment of overseas pilots, the Airline Pilots Association of Singapore (Alpa-S) which represents about 1,600 of SIA's 1,800 pilots, suggested axing the foreign pilots first before accepting a wage cut. These pilots are based in gateway cities to which SIA has the highest traffic, such as London, Sydney and Los Angeles. They are not members of Alpa-S and are not covered by the Alpa-S collective agreement. The union also pointed out that their pay had already been cut when flight allowances plunged due to the reduced number of flights; when management cut pilots' monthly wages by 15 to 22.5 per cent and asked pilots to take 10–12 days of leave without pay every two months which was equivalent to another 20 per cent pay cut. The compulsory unpaid leave was perceived as a deliberate attempt by SIA to enable existing pilots to clear a huge backlog of annual leave through the recruitment of expatriates, thereby causing a high wage cost and a surplus of pilots during SARs. The union stressed that it is not anti-foreign, for about 400 of its members are expatriates, but it is also not foreign-instigated because under the 1981 union charter, expatriates have no voting rights and cannot sit on the executive committee (EXCO). The union was also annoyed when SIA unilaterally replaced the two business class seats with three economy-class seats allocated to pilots for rest. The union also questioned the senior executives' privilege of using the business class for official duty. The disagreement over these issues led the union submit the matter to the Ministry of Manpower (MOM) for conciliation.

In response, SIA insisted that overseas-based foreign pilots are all its employees and as a global company, it avoids any discriminatory labor practices which may result in travelers shunning the airline especially when 80 per cent of SIA income is derived from abroad. The management reassured staff about profit sharing and informed them about the savings of almost S$100,000 a year per pilot by employing pilots based overseas who do not need housing and traveling allowances. As explained, the pay-cut was compelled by the huge wage cost of pilots, the plummeting of profits in 2002–2003 (Figure 7.4) and the extra costs arising from SARs and costs-cutting competition from rivals (*Straits Times*, May 31, 2003). The costs and losses thus wiped out any chance of restoring the proposed 15 per cent across-the-board wage cut. The absence of such unforeseen circumstances and the record profits of S$1.55 billion in fiscal year 2001 made the restoration of the wage-cut possible for the period 2001–2002 and about seven months of bonuses were given out to staff. The years of good profits and bountiful bonuses left workers generally believing that good times will not cease. Managing such workers' unrealistic expectations

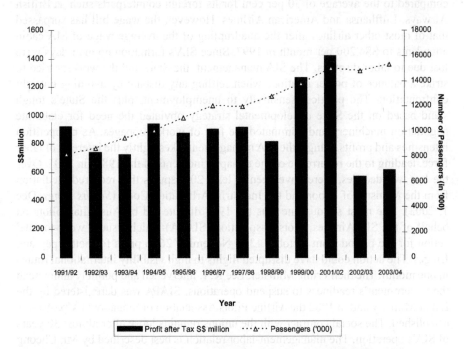

Figure 7.4   Profits after tax and passenger numbers of SIS, 1991–2004

becomes problematic. The Acting Manpower Minister, Labour Chief and Deputy Prime Minister warned the unions of any misconduct in resisting SIA's moves to tide over the bad times. In June, 2003, Alpa-S agreed on the wage cut deal and no-pay leave, but in November, 55 per cent of members voted out the negotiating EXCO team and a deadlock set in (*Straits Times*, Dec 1, 2003). The State was compelled to intervene. Minister Mentor Lee K.Y. stepped in personally 'to finish what he did not in the 1980s', signaling the changing reality and warning the pilots that Singapore's future as a hub was at stake and the importance of keeping its status as part of the national survival strategy. He blamed the pilots' constant absence from the country for their ignorance and for not being part of the NTUC and hence suffering from dislocation in medical payment and retrenchment benefits which the National Wages Council (NWC) had set the guidelines on (*Straits Times*, Feb 27, 2004). NWC is a tripartite body formed by the State with representations from the public sector, labor (NTUC) and designated employers' associations.

Unlike most other Asian countries, SIA started its services by deploying the strategy of high wages and intensive training due to shortage of local labor and skills. Such high costs are factored into the relatively high prices charged to passengers who prefer premium services. The high training costs keep recruitment tightly controlled and in tune with capacity growth. Staff made surplus through technology are retrained, redeployed or induced to leave through generous redundancy schemes. The labor costs of SIA are relatively low (less than 20 per cent of total revenues) compared to the average of 30 per cent for its foreign counterparts such as British Airways, Lufthansa and American Airlines. However, the wage bill has surpassed that of most other airlines after the quadrupling of the average wage of SIA cabin attendants to S$4,200 per month in 1997. Since SIA's formation no man-days were lost due to labor disputes. The SIA management, the State and the workers tried to strike a balance of power relations when settling any dispute by avoiding outright confrontation. The political sensitivity to unemployment, plus the State's tough stand based on the State developmental strategy obviated the need for elaborate negotiation machinery and eliminated the fear of work stoppages. As competition intensifies and profits plunge, the SIA management takes lightly the interests of labor force, leading to the recurrence of the unhappy incidents of the 1980s in 2003. Over the last two decades, there have been at least 20 disputes that required mediation from the Ministry of Labor and the Industrial Arbitration Court (*Straits Times*, Dec 1, 2003). The most serious one was the 1980 dispute led by Australian pilots on behalf of the SIA Airlines Pilots Association (SIAPA) which issued 'work-to-rule' action for the period from October 13 to November 26 to press for better pay and fringes. The action could have disrupted flights to high visibility international routes in one month. The then Prime Minister, Lee K.Y. intervened and publicly announced the government's readiness to suspend operations. SIAPA was deregistered by the Labor Ministry and in 1982 the Airline Pilots Association of Singapore (Alpa-S) was established. The social contract with the unions remains loose after almost 30 years of SIA's operation. The management-labor relation is best described by Mr. Cheong Choong Kong, former CEO of SIA, 'There was confrontation when there should be cooperation and aggressiveness when there should be warmth' (*Sunday Times*, May 3, 1998).

After three months of strained relations, clarification by the management and stern government warnings, in February, 2004, the newly elected President of Alpa-S, Captain Mok Hin Choon announced the union's decision to conclude a new wage deal with SIA by April, 2004. A letter was sent to the MM Lee K.Y., stating the pilots' position on various issues *inter alia* not pressing for pre-SARs salaries, support of a flexible wage structure, not benchmarking pilots' proposed salaries against top or highly paid international pilots, and work with management in private to settle grievances and address slipping morale. In the same month, the CEOs of SIA, SATS, and the SIA Engineering Company met leaders of all five staff unions – Air Transport Executive Staff Union, Alpa-S, SATS Workers' Union, SIA Engineering Company Engineers and Executives Union (SEEU) and SIA Staff Union (SIASU). The union leaders pinpointed the lack of corporate culture, the fear of workers about committing mistakes and retrenchment (*Straits Times*, March 2, 2004). SIA was told to change the divide-and-rule tactics and be more concerned with workers' welfare. They added that middle managers are poor communicators and rarely attempt to explain the organization's decisions or policies, or share information and often refer to instructions from the top as an answer. They suggested measures for improving communication through education talks on the company's financial performance and the business environment, organizing seminars, retreats and social gatherings with different levels of management.

In March, 2004 the SIA's new CEO, Mr. Chew, promised to start relations afresh with the union. He reiterated the management's unchanged stand on employing foreign pilots considering their mobility and high costs of training them. He ruled out the possibility of refunding the 9/11 pay cut at this stage because the existing profit of $631 million was still below SIA's weighted average capital cost of 8 per cent to ensure its long-term viability. The airline accepted the compromise of reserving one space bed and two economy class seats for pilots' rest period and compensating monetarily any pilot who had to sit in economy class. Chew also verified that the privilege of senior executives and board directors using business class when traveling on duty was subject to commercial demand. He explained that the recruitment of expatriates during SARs was to make up for the freezing of recruitment during the 1997/98 crisis as business picked up faster than expected after the crisis. Workers were asked to take unpaid leave because of the accumulation of unconsumed annual leave. To clear the backlog of leave by the end of 2003, additional captains were hired and more flights were assigned to management pilots from March to December 2002 and Alpa-S agreed. The decision was made without contemplating the SARs outbreak in March 2003. After the SARs outbreak, further recruitment was discontinued. The unprecedented reduced number of flights in consequence of SARs enabled the airline to release pilots to clear their backlog of leave. Chew also indicated that there was no large surplus of captains and compulsory unpaid leave was terminated in late 2003. He also revealed that for the period 2002 to 2004, retired pilots and non-flying simulator instructors were hired to create the savings for additional staff to provide better administrative support for pilots (*Straits Times*, March 2, 2004).

In September, 2004, soon after SEEU and SIASU inked a collective agreement with SIA, Alpa-S also signed a new three-year collective agreement with SIA with the following agreements: pay cut of between 11 and 16.5 per cent in 2003 (*Straits Times*, 10 Sept 2004); variable pay increase based on market performance; maximum number of flying hours to increase from 900 to 1,000; annual leave to be extended from 28 to 36 days; the income ratio between the highest earning and the low income captain would be lowered from 1:9 to 1:6. The agreement would help cut labor costs by 20 per cent, saving about $700 million which was still far from the $800 million to $1.6 billion targeted to achieve the 8 per cent return of capital. SIA will convert fixed costs to variable costs for effective capacity management, changing the practice from a price-taker to a price-maker.

Corporate Governance and Triadic Cooperation of the State, SIA and Labour

Despite the disputes, SIA had never faced a strike and adversarial union relations of the scale experienced by USA or UK carriers. As corporations become increasingly global in scope, adhering to legislation of only one country does not suffice to sustain business viability. The ICT convenience today easily allows negative campaigns on a global level. To be vigilant of any rogues in their midst, a global firm would require robust internal control systems and management. The more diverse and international an industry becomes, the more heterogeneous the labor force is, the higher the instability and risk, and the more robust the strategy for internal control. At the same time, employees are becoming increasingly active in the protection of their rights; their sabotage could impact on the turnover and the branded image of a global business like SIA. These are the new corporate governance challenges of the twenty-first century. More research is needed to analyze the vulnerability of a global service that hails from a small, resource-poor city-state as the risks could be higher. The SIA experience indicates that organizations must consistently analyze all risks including workers' dissatisfaction and misunderstanding. A clear, methodological approach to risk identification and management is imperative, particularly for businesses operating on a global scale. Inconsistencies in a global policy will leave an organization vulnerable to negative activist scrutiny. A single incident can destroy a carefully constructed global image. A sound system of corporate governance can exert effective internal control, define clearly the business risks, identify opportunities, streamline decision-making, enhance performance and reduce cost. A good reported performance profile could attract further investment and help keep share prices buoyant. Addressing corporate governance goes beyond ensuring high-level structural requirements. Top-level decisions need to be made with the interests of key stakeholders in mind and the welfare of the workers at heart. An appreciation of global standards should be maintained and communicated to the work force. The sooner internal control and risk management issues are addressed through a sound corporate governance framework the sooner organizations can identify the risks and capitalize on the benefits.

In retrospect the 2003–2004 dispute between SIA management and the pilots union could create the worst fear of unions' hardline approach of calling large-scale strikes, crippling the enterprise's efficiency and income, and most importantly damaging the branded image and a key sector of the economy that speaks of the national survival strategy. The State intervened in timely fashion to nip the problem in the bud by having a dialogue with the union and giving constructive suggestions instead of taking a top-down approach. Why could SIA not be trusted to resolve the problem internally? Probably neither the State nor the pilots' union had complete faith in SIA's management to resolve its labour issues. Another explanation is the lack of any institutional arrangements that can help create the trustworthy and mutually beneficial reciprocal relationship – the weak link between the management and labor. The appointment of NTUC Chief as a member of the SIA Board of Directors since 1997 has its limits. As the majority shareholder of SIA, the State ensures that SIA prioritizes the interest of the country which in turn would require cooperation, trust and confidence from the unions. After the dispute, the government takes a resolute stand. To prevent another such cycle of acrimony, the MOM amended the Trade Unions Act which enables trade union EXCO to negotiate and commit to collective agreements without having to seek ratification from members. MOM will also rescind approval for the two non-citizens who sat on Alpa-S EXCO. The government firmly forbids any industrial strife to disrupt the economy which is already handicapped by the scarcity of land, resources and cost disadvantages. The government reiterated the need of symbiotic relationship between industry, labor and the State for national survival. SIA management was told to undertake wage reform, improve corporate performance especially in human resource management, forge a common understanding with employees and pay competitive wages to retain good staff. Though directed at SIA, the message applies to all companies in the city-state. The strong messages are derived from the 2003 NTUC national delegate's conference when employers were chided for being the weakest link in labor relations, and a motion was tabled urging employers to act responsibly and reprimanded them for failing to look after workers' interests during economic downturn. A recent poll by the Gallup Organization found that about 17 per cent of Singapore employees were unhappy at work, and poor employee management was estimated to cost the economy at least S$4.9 billion in lost productivity (*Business Times*, Dec 2, 2003).

Future Prospects and Concluding Remarks

The past, and recent, episodes of SIA labor disputes show the weak link between industry and labor. The strict labor laws do not help develop the chemistry that is needed to facilitate communication and understanding between the two parties. The developmental state strategy has effectively gained the partnership of labor but could not foster the partnership of industry and labor. Evidently more research is needed in this respect to further investigate the causal factors of such a weak link. In times of crisis, the State has to act as the facilitator, supporter, arbitrator and negotiator as observed in the SIA experience. The State uses competition as a means of control and amplification of benefits. Instead of being protective, the State is pro-business, development oriented and pragmatic, but does not trivialize the organization's financial viability and national interests. By leveraging on the geo-strategic location, the State turns the country into a regional gateway by complementing the services of the airline with other relevant services such as tourism, transshipment and logistics. The smallness of its economy and the city-state characteristic enable it to combine statesmanship with business opportunism. A centralized political system that deliberates decisions from top to bottom allows political leaders to marshal the entire nation's resources in support of any program. The good connectivity to core economies of the world, outsourcing of resources, the constant upgrading with the support of the State and emphasis of quality services have propelled SIA into a branded global service provider.

Using the developmental state strategy to chart the directions of a small city-state as demonstrated by the case of SIA is evidently not sufficient to deal with the many challenges of the new global economy. The case of SIA is unique as its performance is closely intertwined with that of the nation-state and that of the regional and international economy. Both the airline and the country have to grapple with the constraints of the scarcity of resources and expertise; the small domestic market and limited geographic extent which make them dependent on foreign markets and foreign labor. As a global enterprise SIA has to exhibit greater transparency in policy making and consistency in the treatment of its staff. Bridging the gap of understanding of local and foreign labor force and alerting them regarding the importance of their cooperation in sustaining the airline's as well as the nation-state's competitiveness is a formidable task. For any global service provider, particularly one that is based on a small nation-state, the sustainability of competitiveness therefore should not bank on the strong link between the State and labor or between the State and the industry organization *per se*. Winning the trust, understanding and cooperation of labor in a global business transcends the role of the State. Initiative has to come from the industry organization to cultivate the palatable corporate culture that could govern not only its own interests but also that of the national economy. To scale to the level that top USA and EU airlines are flying, SIA needs to move to a new phase in its labor relations. It will be worthwhile to study how corporate governance can help build the atmosphere of trust with the staff through effective communication in addition to raising their productivity, injecting flexibility in recruitment and the wage system. Maintaining good communication with the employees could make all the difference to whether the airline could remain 'A Great Way to Fly' but also become 'A Great Place to Work', which in turn will attract the very human resources that a small nation-state needs to help spur the next stage of development.

To sustain economic growth and development in an increasingly competitive global economy, developing countries compete to attract FDI. The common set of incentives and creation of attractive investment environment give firms a diversity of choices and the mobility, in particular, manufacturing TNCs. The re-location and outsourcing activities of these firms have led to the fear of hollowing out of the local economy, including Singapore. The alarm was raised by the recent massive flow of FDI to China. The service of SIA has been relying on international tourism and business activities, particularly those of foreign TNCs. The fixated infrastructure and location-specific amenities of SIA-Changi make their services less flexible compared to manufacturing, though some of the ancillary services can be outsourced. The labor dispute is manageable with the visible support of the State. To what extent and how the State can do the same effectively for foreign TNCs and tourists? The movements of the two could impact greatly on the overall economy, and specifically their demands for the associated services which a service provider such as SIA is depending on, suggesting a research lacuna that is yet to be explored fully.

References

Bowen John T. and T.R. Leinbach (1995), 'The State and Liberalization: The Airline Industry in the East Asian NICs', *Annals of the Association of American Geographers*. 85 (3), 469–493.

Chan, Daniel (2000), 'Air Wars in Asia: Competitive and Collaborative Strategies and Tactics in Action', *Journal of Management Development*, 19, (6), 473–488.

Chan, Daniel (2000), 'The Story of Singapore Airlines and the Singapore Girl', *Journal of Management Development*, 19 (6), 456–472.

Chang Z.Y., Yeong W.Y. and Loh L. (1996), *The Quest for Global Quality, A Manifestation of Total Quality Management by Singapore Airlines*, Singapore: Addison-Wesley.

Daniels, P.W. (2000), 'Export of Services or Servicing Exports?', *Geografiska Annaler*, 82 (1), 1–15.

Daniels P.W. (2005), 'Services, globalization and the Asia-Pacific region', in Daniels P.W, K.C. Ho and T.A. Hutton (ed.), *Services Industries and Asia-Pacific Cities*, Routledge, New York, 21–51.

Findlay, C., Chia L.S. and Singh K. (1997), *Asia Pacific Air Transport Challenges and Policy Reform*, Singapore: ISEAS.

Ho, K.C. (2000), 'Competing to be Regional Centres: A Multi-agency, Multi-locational Perspective', *Urban Studies*, 37 (12), 2337–2356.

Labich, K. (1994), Air wars over Asia, *Fortune*, 129(7), 93 (4 pages).

SIA, *SIA Annual Report*, Singapore.Various years.

Wong Shuang Yann (2005a), 'Cross-National Ethnic Networks in Financial Services: A Case Study of Local Banks in Singapore', in Claes G. Alvstam and Eike W. Schamp (ed.), *Linking Industries across the World*, Ashgate: England, pp. 243–276.

Wong Shuang Yann (2000b), 'State Governance, regulatory Processes and Entrepreneurship: Singapore's Concentrating Banking Sector', in Richard Le Heron and James W Harrington (ed.), *New Economic Spaces: New Economic Geographies*, Ashgate: England, pp. 73–82.

Yeung, H.W.C., Poon J. and Perry M. (2001), 'Towards a regional strategy: the role of regional headquarters of foreign firms in Singapore', *Urban Studies*, 38 (1), 157–83.

Chapter 8

Integrating Foreign-Owned Firms into the Global Value Chains: A Case Study of Window Korea Project in China[1]*

Shuguang Liu** and Guogang Ren***

Introduction

Although the concept of a value chain (Porter, 1985) may refer to economic interactions at any scale, one of the major applications of value chain analyses is to the global expansion of multinational companies (MNCs). The categorization of global value chains (GVC) into producer-driven and buyer-driven by Gereffi and Korseniewicz (1994) is also based on an analysis of the worldwide activities of MNCs. It is not difficult to realize that GVC has become the tool for MNCs to establish global governance orders in diversified ways (Humphrey and Schmitz, 2000). At the same time, much has been discussed about the way in which, and to what extent, MNC-dominated global chaining affects local clusters and even the local economy as a whole. Local industrial upgrading along the global value chain has become the hot topic, especially in developing countries.

As for the responses and contributions of small and medium-sized enterprises (SMEs) to global value chains, more alternative has been given since the renaissance Marshallian industrial district theory by Italian researchers such as Bagnasco (1977), Garofoli (1978) and Becattini (1979). Park and Markusen (1995) consider that the development of new industrial districts is international trade oriented. Markusen (1996) classifies new industrial zones into 4 types of Marshallian, hub-and-spoke, satellite platform and state-anchored, indicating the unique roles of local SMEs in the globalization of economic activities. Kaplinsky and Morris (2000) summarize the possible ways in which SMEs participate in GVC as (1) lobbying government for

---

1 *The authors acknowledge the financial support from Ministry of Science and Technology of China for the research conducted (research grant No. 2003DGQ2D059).
** School of Economics, Ocean University of China email: dawnliu9631@263.net.
*** Beijing Research Institute, ZGW Strategic Consultancy email: net806@21cn.com

assistance and (2) undertaking joint activities, such as quality auditing or, branding, especially with regard to learning-networks.

There are also considerations in regard to the consequences of adopting MNCs for local SMEs. From the limited literature we can find that the offshore tendency of MNCs pursuing global strategies are very strong in post-industrial countries like Japan; sometimes they try to achieve local embeddedness at the expense of remaining detached from rest of the MNCs (Lehrer and Asakawa, 2002). The outward expansion of MNCs will bring some shocks for local SMEs, but will help their upgrading in the long run. Head and Ries (2002), for example, find a positive influence of offshore production by Japanese multinationals on domestic skill intensity.

With worldwide industrial restructuring, the advanced countries are upgrading their industries toward knowledge-intensive services while manufacturing activities are dispersing into developing countries like China. Not only MNCs have managed to establish their strategic value networks in developing countries, more and more SMEs have began their processes of cross-border transfer as well. In China, more overseas SMEs have mushroomed in recent years, but many of these firms have to face up to an alien local milieu, and there are more SMEs facing the dilemma of stay at home or go abroad. On the other end, local governments in China are enthusiastic to host and serve the domestic and foreign enterprises by establishing many more business incubators, but the results are not very satisfactory. The purpose of this chapter is to find possible ways for overseas SMEs, especially the improvement of international business incubators, to participate in global value chains in the local milieu of China.

The Patterns of Overseas Enterprises Survival and Growth in China: Is the Business Incubator Necessary?

The Patterns of MNCs Value Chain Networking Strategy

The development of overseas MNCs in China can be divided into three stages. Before the 1990s they managed to establish selling agents/companies in Hong Kong and southern part of Mainland China. Since the early 1990s more and more MNCs began to establish production units in China. At first they take the form of 50:50 joint ventures but they owned a higher percentage afterwards. With the impending entry to the WTO, the competition among MNCs and even domestic firms became severe in China, and MNCs began their strategic readjustment through integration of already existing branches and the establishment of regional headquarters, R&D, as well as training centers.

Case 1: Strategic development of Siemens in China. Siemens began cooperation with China as early as 1872; the year Siemens supplied the first pointer telegraph to China. To date, Siemens has established more than 40 companies, 28 regional offices and has over 21,000 employees in China. Siemens takes the strategies of

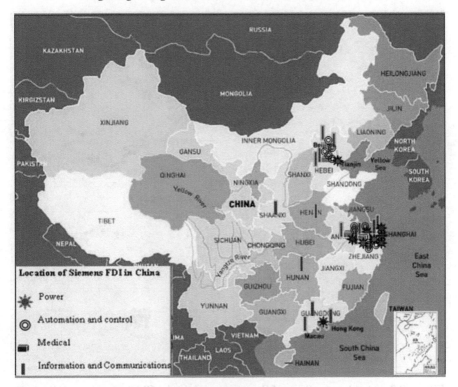

Figure 8.1   The location of Siemens offices in China

cooperation and location respectively under the common brand and integrated top-level decisions. It seems that the Siemens companies in different industries have concentrated in special regions, but in fact their local interactions are not very strong (Figure 8.1).

Case 2: Sanyo agglomeration in China. Sanyo established a presence in Beijing in 1979, but it developed its businesses in limited areas of the Pearl River Area at Shenzhen, Dongguan, Foshan and Guangzhou since the 1990s. Sanyo has also expanded northward in Dalian and Shenyang, as well as competing with MNCs from the USA and Europe in Yangtze River Delta region, e.g. Shanghai and Suzhou. Sanyo began its integration action by establishing Sanyo Electronics (China) LTD in Beijing in 1995, aiming at providing services in finance and accounting, legal affairs, information survey as well as marketing. The major forms of Sanyo value chain activities can be summarized as: (1) at the initial stage it manufactures in mainland China and sells to worldwide markets via Hong Kong; (2) gradually it focuses on manufactures and sells within China; (3) it begins to seek cooperation with domestic partners and oversees the branches of overseas MNCs in China; (4)

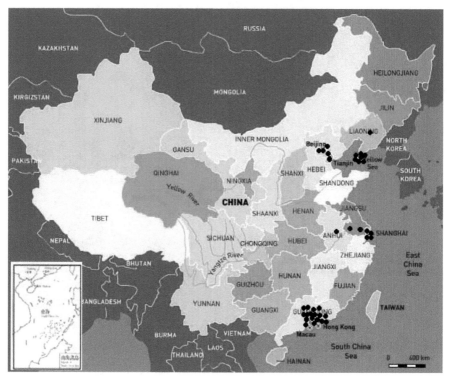

Figure 8.2   Agglomerations of Sanyo companies in China

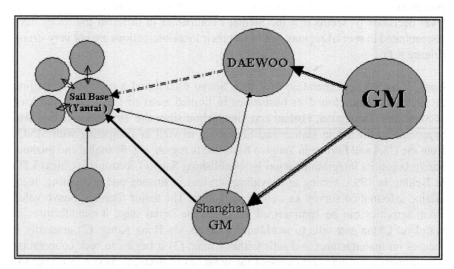

Figure 8.3   GM value chain expansion in China: a case of Sail Base

Sanyo companies have formed relatively closer relation in special places such as Pearl River Area.

Case 3: GM value chain expansion in China. Although General Motors is a latecomer in China's car market, it has made rapid progresses and began its rapid value chain expansion processes. By taking over Shandong Yantai Auto Corporation, GM (China) and GM Shanghai realized the successful transfer of Buick (and now Chevrolet) lower class sedan car manufacturing base from Shanghai to Shandong, and begin to form the car parts subsidiary system through GM Daewoo and its former branches in Qingdao and Weihai.

Overseas SMEs Survival in China

Case 4: Taiwan IT SMEs cluster in Dongguan and Kunshan. The location of enterprises from Taiwan is relatively concentrated in Guangdong, Jiangsu (including Shanghai), Fujian, Hebei (including Beijing and Tianjin), and Zhejiang. The Taiwan IT SMEs have formed clusters in the Yangtze River Delta and the Pearl River Delta (Wang and Zheng, 2002). The major features of the Taiwan IT cluster are: (1) they form integrated production systems mainly through value chain investment in limited areas; (2) they keep global value chains all the time, and become the core part of global IT value chain, but most of them are at the lower ends of the value chain; (3) they are organized mainly through industrial associations already established in Taiwan; (4) they have established close relations with local government.

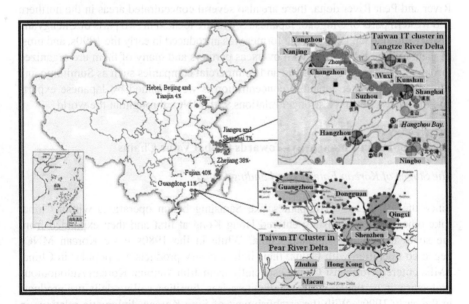

Figure 8.4   Taiwan IT clusters in Mainland China

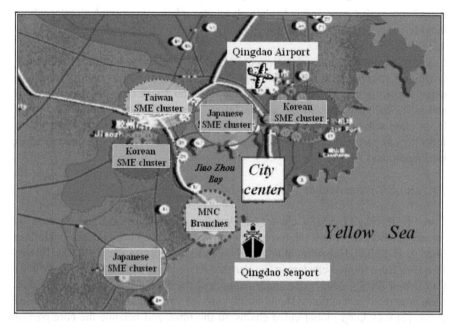

Figure 8.5   East Asia SMEs in Qingdao, Shandong province

Case 5: East Asian SMEs clusters in Qingdao. Although the agglomeration of overseas SMEs in Qingdao, Shandong province is not as strong as in the Yangtze River and Pear River delta, there are also several concentrated areas in the northern and western parts of Qingdao. Japanese SMEs are located in two parts of Chengyang district and Jiaonan District. The former was introduced in early the 1990s, and most of them engage in IT OEM with overseas partners and many of them are organized through large Japanese industrial and commercial companies such as Sumitomo and Mitsubishi. The later Japanese concentration is well planned by Japanese experts and the enterprises have stronger relations with MNCs throughout the world.

Korean SMEs in China: Ways Towards Global Value Chains

The Stages of Korean Enterprises Landing China

Since the 1970s large companies like Samsung begun operations within China. Like other MNCs, Samsung entered Hong Kong at first and then expanded from the south to rest of Mainland China. While in the 1980s more Korean MNCs began co-operation with China, their light industry products are popular in China. SMEs entered Northeast China, especially from Jilin Yanbian Korean Autonomous prefecture, in particular really small enterprises, families, selling daily use products in the early 1990s. With the establishment of Sino-Korean diplomatic relations in

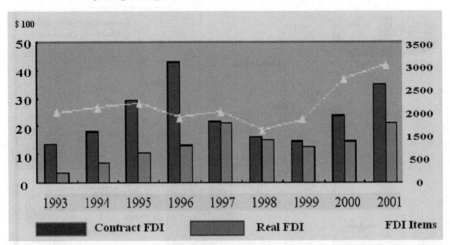

Figure 8.6   The process of Korean outwards FDI to China

1994, Korean FDI to China increased gradually, but decreased a lot during the Asian financial crisis. From 2000, Korean MNCs begin to adjust their strategies in China as they began to compete with other MNCs giants, and try to win the attentions of customers in rapidly changing markets. With the embedding of Korean MNCs in China, more and more Korean SMEs follow up to seek opportunities not only with their former partners, but Chinese partners and even other overseas partners in China (Figure 8.6).

Reasons for Industry Transfer to China

According to recent research, the motivations for Korean FDI to China are as follows (Liu and Yang, 2004): (1) Domestic economic re-construction toward knowledge-intensive and service industries, to some extent driving out the traditional industries; (2) The rapidly rising cost of raw materials and labor has made the Korean enterprises less competitive in formerly competitive industries such as clothes, consumer electronics, foods, etc; (3) It allows Korean enterprises to change the situation of negative competition with USA and European corporations in China, most of them have accomplished their value chain networking in China, leaving limited room for Korean firms; (4) The large market potential of China has changed the traditional ways of early time Korean enterprises who just utilized cheap labor and land in China. More Korean enterprises attempt to open and expand Chinese market, so it appears more important to establish manufacturing units in China; (5) The improvement of manufacturing activities in China, as well as in Shandong province, has made it easier for Korean investors to find suitable partners and suppliers.

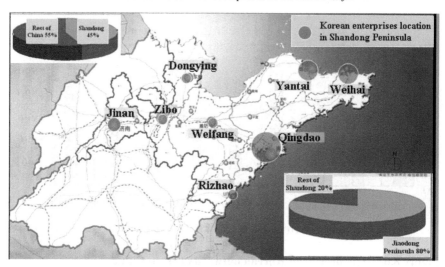

Figure 8.7   Location of Korean enterprise in Shandong province

Spatial Concentration of Korean Enterprises in Shandong Peninsula

With the closer Sino-Korean economic relations since the late 1990s, China has attracted more and more Korean FDI, with 45 per cent of the total investment concentrated in Shandong Province. The Korean enterprises are concentrated in special areas in Shandong, more than 80 per cent located in eastern part of Qingdao, Yantai and Weihai. An indepth interview of some Korean SMEs in Shandong, China and Incheon, Daegu, Pusan, Korea, was carried out. The question was very simple: (1) Why do you choose Shandong or why not? (2) What kind of help do you need for investing in China/Shandong, is the international business incubator necessary for you?

We list the results according to the frequency of the answers on major factors of Korean SMEs agglomeration in Shandong peninsula: (1) lower cost of transportation of cargo and passengers based on geographical vicinity; (2) similarity of natural environments; (3) cultural and behavioral compatibility; (4) localized Korean networks for latecomers; (5) technical and industrial match; (6) already existing market chains; (7) special services provided by local government; (8) fewer competition pressures than other area; (9) easiness of finding Korean language speaking employees.

As for the major reasons for not coming to China/Shandong peninsula: (1) strange to Chinese cultural and social environment; (2) language barriers and difficulties in seeking satisfied employees; (3) difficulties in financing; (4) difficulties in local production networking; (5) difficulties in accommodations and entertainment.

As for the question of whether or not special business incubators are necessary, most said yes, but what kind of incubators should be provided and how they operate is still an open question, for there are few successful cases.

Evaluation of International Business Incubators in China

The Concept of Business Incubators

Business incubation can have several definitions and approaches. According to the National Business Incubators Association:

> Business incubation catalyzes the process of starting and growing companies. A proven model, it provides entrepreneurs with the expertise, networks and tools they need to make their ventures successful. Incubation programs diversify economies, commercialize technologies, create jobs and build wealth.

according to the NBIA, the major function of business incubators are to:[2]

- *Make a difference in their communities.* In 2001 alone, North American incubators assisted more than 35,000 start-up companies that provided full-time employment for nearly 82,000 workers and generated annual earnings of more than $7 billion.
- *Create successful companies and reduce the risk of investment.* Startup firms served by NBIA member incubators annually increased sales by $240,000 each and added an average of 3.7 full- and part-time jobs per firm. Business incubators reduce the risk of small business failures. NBIA member incubators report that 87 per cent of all firms that graduated from their incubators are still in business.
- *Create economic development opportunities.* For every $1 of estimated annual public operating subsidy provided the incubator, clients and graduates of NBIA member incubators generate approximately $45 in local tax revenue alone. NBIA members report that 84 per cent of incubator graduates stay in their communities and continue to provide a return to their investors (NBIA, 1998).

Classification of Business Incubators in China

According to the differentiation of investors, structure, and the functions, we classify six kinds of business incubators as follows:

- Comprehensive incubators: they try to help start-ups confronted with the most common problems that require office space, help with company registration,

---

2 Source: www.nbia.org; www.21sh.com/biti/about/usa.htm.

providing preferential policies, etc.
- Incubators for start-ups in special technology development. They aim at incubator firms with special technologies such as biology, software, new materials.
- Incubators for special talents. Government invests most in those incubators that are aimed at attracting overseas talent to invest and develop in China.
- International enterprise incubator. The objective of these incubators is to attract foreign enterprises to develop in China and domestic enterprises to develop abroad.
- Virtual incubator. This kind of incubator provides incubating services via the Internet.
- Enterprises invested incubator. This kind of incubator is always established by the large enterprises in preparing for new projects and risk-taking development activities. For instance, some of the world's top biotech enterprises have established several special incubators in Pudong region, Shanghai.

Development and Location of Incubators in China

China initiated its incubation system with the establishment of Wuhan Donghu Incubator in 1987. Nearly 30 incubators were set up between 1989 and 1990. The incubators, especially the technology incubators, developed quickly as more tenants graduated from them (Figure 8.8). According to the report on China Incubator Development (2003), the development of incubators in China can be divided into two stages, the classical period and the diversified period. Government invests heavily in the construction of incubator infrastructures and therefore improves the conditions for start-ups. The second stage started from the late 1990s.

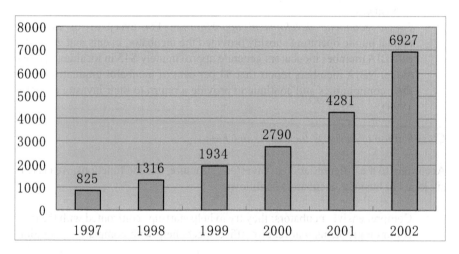

Figure 8.8   The growth of technology incubators in China (1997–2002)

The main feature can be summarized as follows: (1) they try to provider wider range of services to start-ups, including help with marketing and seeking employees; (2) rapid development of special incubators, especially biotech incubators and software incubators; (3) Establishment of venture capital; (4) establishment of R&D and market networks with relevant groups; (5) try to earn profit through running incubators; (6) development of international business incubators in the form of overseas talent incubation, international networking.

The government-supported incubators are mainly concentrated in the development zones of large cities (see Table 8.1) but always far away from the location of overseas SMEs clusters.

Incubators in China: The Current Picture?

Although no one could ignore the progress of incubators and the contributions to start-up, the problems are also obvious in relation to the high quality services, especially in technology industrialization and venture capital provision. Sometimes the incubators become a refuge for small firms that wish avoid to severe competition, while some other SMEs that really need help do not get in. Just as Lakaikai (1999) remarks:

> The program is heavily focused on the 'hardware' aspects of incubators. Importance of soft services is only now being recognized. The program has not been immune from the dynamics of politics. Local 'empire building' is an important driver of the program. This skews the effectiveness of investment in the program ... and the relative neglect of services. Incubator managements are generally composed of civil servants who have little or no entrepreneurial experience. This further limits the quality of the 'soft' business support services they can provide to their tenants. The services that are provided in-house are typically not on a cost-recovery basis. This limits their quality and sustainability.

Table 8.1   The location of incubators in cities

Place	Incubators	Place	Incubators
Anshan	229	Shanghai Yangpu	225
Beijing Zhonguancun	248	Shanghai Zhangjiang	280
Chengdu	289	Shenyang	249
Chongqing	225	Shenzhen Shanqu	268
Dalian	279	Xian	200
Fudan	211	Xian	230
Harbin	240	Zhengzhou	260
Hefei	238	Suzhou	320
Jilin	306	Wuxi	345
Luoyang	215	Shenzhen	385

Source: China Incubator Report, 2003.

There is as yet no coherent national policy framework for incubator development. Few guidelines in terms of feasibility study templates, operations manuals, evaluation criteria, etc., exist. As in some other countries, incubators are launched and operated primarily on the basis of 'intuition.'

As is outlined earlier, overseas SMEs need comprehensive help that should be available via the incubators. But in reality, the incubators are always domestically oriented, and their functions are determined before the actual needs of SMEs. The already existing incubators are mainly located in large cities; there are spatial mismatches between SMEs and business incubators. On the other hand, many SMEs clusters have began to rescue themselves through local and global networking by resorting to their own associations and/or traditional linkages.

Case Study of Window Korea Project

The purpose of this project is to help Korean SMEs to establish in the Chinese market and the local milieu, therefore improving the success rates for survival, growth, and even transferring to ideal locations. This project takes the model of PPP (Public and Private Partnership) between Wendeng local government and Shengji Co. Ltd. The location of the project is at Weihai at the eastern end of Shandong province, the nearest location with Korea.

The project invited the consulting firm ZGW Strategic Consultancy, a famous private consulting company in China, to develop initial decision-making and management guidance. Based on the general analyses of new trends in international industrial transfer, relationships of Northeast Asia, and the decisions and actions of central and local government, the project is initially committed to establishing Korean SMEs business incubators in China under the banner of Window Korea in China.

As the actual initiatives of the project, Shanghai Modern Group, China and Korean Space Group in Seoul, Korea, have worked out a master plan through close collaboration. And all the process is based on the guidance provided by ZGW Strategic Consultancy. The master plan shows comprehensive functional units such as the business and financing center, exhibition and distribution center, training and education center, Korean traditional cultural center, as well as residential districts (Figure 8.9). Dynamic plans are also worked out for realizing the possible requirements of Korean SMEs tenants in global value chaining in the future.

Conclusion

Accompanying the decentralization process of MNCs, more SMEs have begun their process of transferring to other countries. The survival patterns of SMEs moving to an alien milieu are different to MNCs; they prefer to seek easy locations and to establish their local networks largely through their own efforts. Although many SMEs (like the case of Taiwan IT clusters in Dongguan and Kunshan) transfer their

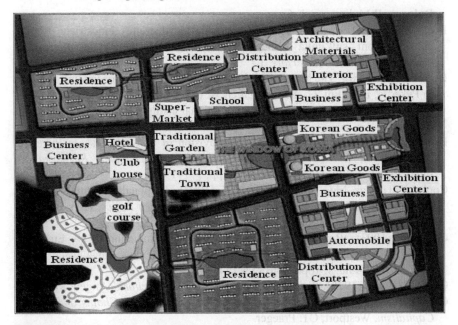

Figure 8.9   Conceptual design map of Window Korea

locations without losing their global value chains, there are still some that have to suffer from the difficulties in joining into global value chains in alien milieu. They need help to improve their survival and upgrading. Business incubators have been introduced to help start-ups and international business incubators should play a role in helping SMEs with international backgrounds, no matter whether they are domestic or overseas. But the development of government-driven incubators in China has been in trouble to meet the needs of overseas SMEs very well. There are mismatches not only of space, but functions as well. The Window Korea project tries to probe into the new ways of helping overseas SMEs through establishing the PPP model international business incubators. Although many Korean enterprises and SME associations have indicated their willingness to participate in the project and become members, more issues than actual plans and constructions remain to be addressed.

References

Amin, A., Thrift, N. (1994), *Globalization, Institutions, and Regional Development in Europe*, Oxford: Oxford University Press, pp. 14–15.

Aydalot, P. (1986), *Milieux innovateurs en Europe*, Paris: GREMI, pp. 5–28.

Bagnasco, A. (1977) *Tre Italie: La Problematica Territoriale dello Sviluppo Italiano*,

Bologna: II Mulino, pp. 1–6.

Becattini, G., (1979), Dal "settore" industriale al "distretto" industriale. Alcune considerazioni sull'unità d'indagine dell'economia industriale, *Rivista di economia e politica industriale*, 1.

Becattini, G. (1990), 'The Marshallian industrial district as a socioeconomic notion', in Pyke, F., Becattini, G., Sengenberger, W. (eds), *Industrial Districts and Inter-firm Co-operation in Italy*, Geneva: International institute for Labour Studies, pp. 31–57.

Brusco, S. (1992), 'Small firms and the provision of real services', in Pyke, F., Sengenberger, W.W., *Industrial Districts and Local Economic Regeneration*, International institute for labour studies, pp. 177–196.

Cooke, P., Uranga, M.G., and Etxebarria, G. (1997), 'Regional innovation systems: Institutional and Organizational Dimensions', *Research Policy*, (26): pp. 475–491.

Cooke, P. (1993), 'Regional innovation systems: an evaluation of six European cases', in Getimis, P. and Kafkalas, G., *Urban and Regional Development in the New Europe*, Athens: Topos, pp. 133–154.

Garofoli G..(ed) (1978), *Ristrutturazione Industriale e Territorio*, Angeli: Milan.

Gereffi, G. and Korzeniewicz, M. (eds.) (1994), *Commodity Chains and Global Capitalism*, Westport, CT: Praeger.

Head, K. and Ries, J. (2002), Offshore Production and Skill Upgrading by Japanese Manufacturing Firms, *Journal of International Economics*, 58, 81–105.

Humphrey, J. and Schmitz, H. (2000), Governance in global value chains. *IDS Bulletin*, 32 (3), 1–14.

Kaplinsky, R. and Morris, M. (2000), How do South African Firms Respond to Trade Policy Reform? In Jalilian, H., Tribe, M. and Weiss, J. (eds) *Industrial Development and Policy in South Africa*. Cheltenham, Elgar.

Lehrer, M. and Asakawa, K. (2002), Offshore knowledge incubation: The third path for embedding R&D labs in foreign systems of innovation, *Journal of World Business*, 37(4), 297–306.

Liu, Shu-guang and Yang, Hua (2002) 'Progress in global value chain and regional industrial upgrading', *Journal of Ocean University of China*, (2): pp. 51–56.

Liu, Shu-guang (2004), *Regional Innovation System*, Qingdao: Ocean University of China Press.

Markusen, A., Lee, Y-S. and Digiovanna, S. (1999), *Second Tier Cities*, Minneapolis, MN: University of Minnesota Press, pp. 3–19.

Markusen, A. (1996) 'Sticky places in slippery space: a typology of industrial districts', *Economic Geography*, 72 (3): pp. 293–313.

Marshall, A. (1910), *Elements of Economics of Industry*, London: Macmillan, pp. 19–48.

Marshall, A. (1961), *Principles of economics*, Cambridge: Cambridge University Press (First published in 1890), 15–20.

Park, S.O. and Markusen, A. (1995), 'Generalizing new industrial districts: a theoretical agenda and an application from a non-Western economy', *Environment*

  and Planning A, 27 (1): 84–104.
Piore, M. and Sable, C. (1984), *The second Industrial Divide*, New York: Basic
  Brooks, pp. 2–11.
Porter, M. (1985), *Competitive Advantage*, New York: Free Press.
Porter, M.E. (1998), 'Clusters and the new economics of competition', *Harvard
  Business Review*, 76 (6): 77–90.
Scott, A. (1988), *New Industrial Space*, London: Pion, pp. 1–28.
Storper, M. (1997), *The Regional World: Territorial Development in a Global
  Economy*, New York: Guilford Press, pp. 19–22.
Wang and Zheng, (2002), 'Decisions of Taiwan investment cluster in China',
  Economic Issues, p. 10.

Chapter 9

Transformation of an Industry Stimulated by Local Economic Growth Policy: The Case of the SHOCHU (Liquor) Industry in Japan

Atsuhiko Takeuchi, Hideo Mori and Koshi Hachikubo

Introduction

An effective regional industrial policy should be grounded in the physical, historical, and socio-cultural conditions of the designated area. This will make it possible to create new types of industry based on a variety of local resources and to promote continual transformation of production systems. Moreover, it will be an actor to encourage local-community-oriented marketing and services that will change economic conditions. Basically, since the 1960s economic policy in Japan has been designed to promote the dispersion of industry ranging from the metropolitan regions to the provincial areas. However, despite huge investment, there have been no drastic changes in the regional economic system. In such a situation, the 'one-village/one-product' (OVOP) project has been introduced to stimulate local economic growth by taking advantage of local resources together with the latest potential in the area. It is a policy originally guided by local government that started in the 1980s, and has resulted in the enhanced development of the local economy. The example of the development of SHOCHU[1] industry is used to illustrate this policy. What are the attractive features of the OVOP? How has the SHOCHU industry been developed and transformed as the result of its engagement with the project? In this chapter, we will investigate, firstly the concepts involved in the OVOP, secondly, the formation and development of the SHOCHU industry, and thirdly the process of transformation of the SHOCHU industry as a result of continual improvements in its production system in spite of the strong pressure in Japan and elsewhere from the traders and associations representing Scotch whisky.

---

1 SHOCHU is a form of liquor.

Regional Industrial policy of Japan and 'One-Village/One-Product Policy'

Regional Industrial Policy by the Central Government

The foundation of Japanese regional economic policy after the Second World War has strongly been based on the dispersion of manufacturing plants accumulated in the three metropolitan areas (Tokyo, Osaka, and Nagoya) to the provincial areas. The central government's aim was to attain a homogeneous development of national land, using industry as a pillar. Projects such as 'New industrial cities' in the 1970s and 'Technopolis' in the 1980s were designated as representative examples (Takeuchi, 1995). The Japanese economy ceased its growth in the 1990s and the economy began a process of drastic restructuring. In such circumstances, the Japanese economy was faced with the challenge of not only strengthening its international competitiveness but also managing the transfer to information-oriented systems under the service economy. For this reason, the central government currently plans to disperse offices that are unexpectedly concentrated in the metropolises to the provincial areas in close aligninment with nationwide information networks. However, no drastic change has been effected thus far due to the fact that the foundation of regional industrial policy is dependant upon activation of industry in the provincial area. In the context of the central government's industrial dispersion policy, the pressure applied by the Diet (the country's legislature) members representing constituencies in provincial area, has been considerable. This fact is a common feature of all the regional policies in Japan. For example, although only five or six districts were originally designated as 'the new industrial city' or 'the Technopolis', more than 20 districts were actually designated at the final stage as the result of the strong pressure applied by Diet members.

As a result of the strong dispersion policy accompanied by long-standing large-scale investment, a large number of plants including high-tech sector have been dispersed actually in all over the country (Takeuchi, 1994). However deducing from the high growth of the Japanese industry that continued through the 1980s, it is commonly assumed that industry can be easily dispersed throughout the country influenced by the guidance of influential economists. But, judging from the detailed research by economic geographers on the actual dispersion of the plants, it is understood that plants whose management is dependent on low-wage and relatively low skill labor account for almost all the industry. Except for several strategic or pivotal plants of major enterprises, almost no tight industrial complexes have been formed (Takeuchi, 1996). In the case of Ohita prefecture (discussed later), for example, large scale plants of iron and steel, petro-chemical, micro-electronics and so on chose locations by the determined projects of central government; more than 60 per cent of industrial production in this prefecture are from these sectors. However, most of them are only the production base of the enterprises, there are no attached R&D or national command and control function. That is to say, almost no change has been achieved in nationwide industrial system concentrated in the metropolitan areas such as Tokyo where giant industrial complexes comprise a variety of establishments

(Takeuchi and Mori, 2001). As a matter of course, none of the recent dispersion has involved control functions found in metropolises, especially of Tokyo.

One-Village/One-Product Project

The regional industrial policy of the central government to disperse industry from the metropolitan areas to the provincial areas and to attain the economic growth of the latter has failed despite huge investment. In such a situation, One Village/One Product (OVOP) started by a local government in Kyusyu has been successful in developing industry by taking advantage of local conditions and the promotion of the regional economy. In this context, attention should be paid to the OVOP project. This project is far from conspicuous because it has been hidden within the large-scale policies of the central government. It was proposed by Mr. Hiramatsu, governor of Ohita Prefecture, Kyushu, and started in the 1980s. Mr. Hiramatsu, who was a supervisor at the Ministry of International Trade and Industry (MITI), had long been concerned with economic development of the provincial region, and had made public a variety of his proposals. After he became the governor, he promoted the OVOP project as a pillar of his policy. He enjoyed overwhelming popular support and this continued even after his retirement. A fundamental concept of the OVOP is independent innovation at the local level, i.e. people living in individual cities, towns and villages should be willing to engage in innovation by themselves and encouraged to produce original products making great use of local resources. The OVOP project allows all the local areas to produce at least one product or project, and furthermore allows them to market the products throughout Japan as well as overseas. Thus, to assist these grass-root projects, the prefecture government was willing to provide various technical support and aid for marketing for their business promotion. The fundamental purpose of the policy was not only to enhance the mind, pride, and self-dependence of local people but also with to nurture local leaders. This is highly regarded as a policy to be continuously propagated on an international scale aiming at the development of local economy together with social and cultural expansion. Furthermore the project is characterized by the prefecture government as triggering business setups that stimulate the economy on behalf of the central government.

In Ohita Prefecture, when the OVOP project started in 1980, 140 products were identified. Among these products, four attained sales of more than 1 billion yen. Currently, the number has risen to 312 and 18 of these have sales of more than 1 billion yen. Agricultural (vegetables, flower, fruits and so on), forest, and marine products and outputs obtained by processing these products (such as pickles, hams and so on) account for almost all of the products. Some service industries associated with traditional events and parks are also included in the project. The OVOP project spread from Ohita to adjacent prefectures in Kyushu, and furthermore extended to other provincial areas throughout Japan. In contrast to the conventional policies of the central government, the OVOP project also contributed more to social and cultural development in the local areas. The principle of the OVOP project is focused first,

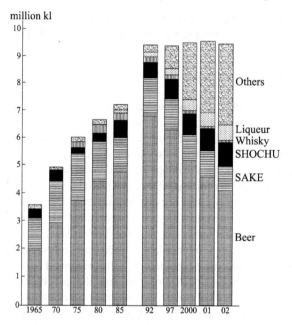

Figure 9.1   Consumption of alcoholic drink in Japan

on operation at the local level but recognizing the global; second, on self-reliance and creatively; and, third, on human resource development. The people living in the individual cities, towns and villages should make efforts to build up a business to produce original products, making extensive use of local resources, supported by marketing throughout the whole country and overseas. Local communities are provided with pride and self-worth by introduced a specific product as a symbol of local identity while the project also plays a role in reversing population decline in agricultural villages. A good example is provided by SHOCHU which has been influential in marketing the application of the OVOP to nationwide industrial systems. Its importance as a demonstration project will be explained below.

Expansion of the SHOCHU Industry

Production and Consumption of Alcoholic Drinks in Japan

Figures on the consumption of alcoholic drinks in Japan show that beer, which accounts for 45 per cent of the total, overwhelms other drinks. SAKE and SHOCHU are distant second (Figure 9.1).[2] On the other hand, wine and whisky are ranked

---

2   Others are composed mainly of beer-like alocoholic beverage (HAPPOUSHU) and various liqueurs.

very low. Before the Second World War, SAKE overwhelmed the other drinks in its ranking, and beer production ran to about half of SAKE production, whereas SHOCHU accounted for just one third of SAKE. After the war, beer production drastically increased, exhibiting an overwhelming share from the 1950s. For the first time in 2001, SHOCHU production exceeded that of SAKE. Meanwhile, wine and whisky (which are being imported) are equivalent to less than 10 per cent of SHOCHU and SAKE in the production. It should be noted that, there is a great difference between SAKE and SHOCHU. The former is utilized as the sacred dedicatory object for New Years Shinto ceremonies and for various traditional events and rituals. The latter is never used as a substitute for SAKE on such occasions. For this reason, sales of SAKE remain steady, supported by deep-rooted tradition in Japanese society.

When analyzing the production and consumption of alcoholic drinks in Japan, it is necessary to take into account the people's drinking habits. The usual style of drinking is different to that of Western or Chinese custom. At dinner or supper in Japan, drinking is usually separated from eating and some dishes are especially prepared for the drinking. One way of drinking in Japan is to continuously consume the drink from the outset of the supper or dinner. Another is to switch to different drinks. The latter is a popular style among the Japanese. This is reflected in the consumption. It remains to be seen what kind of drink will be chosen in combination with beer. In general, SAKE has been popular as a drink after beer, but the number of people who drink SHOCHU after beer has been increasing rapidly in recent years. Beer has enjoyed the overwhelming share of sales as it is the drink that is most commonly selected as the 'first' dinner drink. Wine is consumed largely in combination with western dishes. Whisky is mostly drunk by being added and mixed with ice, water or soda and mostly at parties.

Production Areas of SHOCHU

There are two types of SHOCHU. Type A is produced using a single-step distillation method using several kinds of plants such as wheat (52 per cent), rice (19 per cent), sweet potato (15 per cent), corn, potato, back wheat, chestnut and so on. There are some 300 SHOCHU makers using this method and there is considerable variation in the taste of type A. Type B is obtained by pouring water into highly-purified alcohol enriched by continuous distillation. There is not so much difference in the flavor of type B by the brand. Production of type A is subject to seasonal factors and it is seldom stored for more than one year. This helps to distinguished SHOCHU with other spirits such as whisky because it is fresh. Large scale production of type A is concentrated in Kyusyu in places such as Ohita, Kagoshima, Miyazaki and Kumamoto prefecture (Figure 9.2). The production areas were originally set up and expanded with local markets as its main objectives. On the other hand, the production areas for type B are located in the metropolitan areas. Because of the some major makers of type B engage in large-scale production directed toward nationwide markets. In this paper, the subject of our investigation is mainly type A.

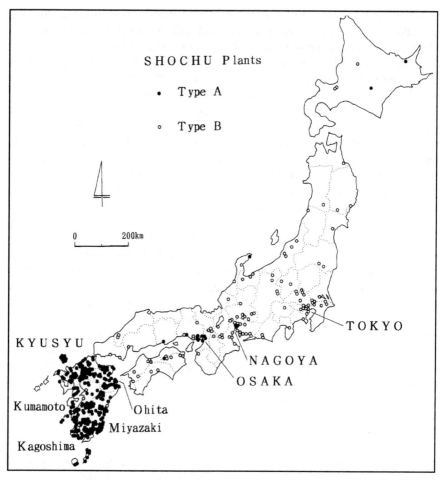

Figure 9.2   Distribution of SHOCHU Plants (2002)

With drinks other than SHOCHU, makers deep-rooted in local markets (as seen in beer production in Germany) have successfully been in operation in all over the country for more than 100 years ago (Hachikubo and Oda, 2000). Distillers of well-known brands have concentrated in Kyoto, Kobe, and elsewhere but their market share is quite small. SAKE makers have played an important role in the local economy in most of the provincial areas. Beer production has been monopolized by four major brewers such as Asahi, Kirin, Sapporo, and Suntory. In the case of whisky, two major makers such as Suntory and Nikka account for almost all of the domestic production, but the proportion of imports is high because Japanese people have a deep-rooted scotch preference and it is necessary for the maker to import original malts from Scotland to produce high quality whisky.

Expansion of SHOCHU Production

Type A SHOCHU, which has been in existence for a very long time as a popular drink in Kyushu, has never spread throughout the country. On the contrary, type B, which has been regarded as an inexpensive and low quality alcoholic drink, has been consumed very widely in the metropolitan areas since before the war. That is to say because of its low cost, SHOCHU is firmly accepted as a drink for lower income groups and the unemployed. As a matter of course, SHOCHU was neither drunk in high-class restaurants nor sold in department stores. Just after the second world war, as most Japanese were very poor, cheap and low quality mass produced SHOCHU produced by mass production was quite a popular drink among laborers or students, and its production made remarkable growth. Whisky that spread as well at that time was a water-diluted drink made from inexpensive whisky. During the 1960s, the Japanese economy grew rapidly, and incomes also increased considerably. At first, the consumption of SAKE increased, but beer production increased at a faster rate. On the other hand, the production of SHOCHU remained stagnant relative to competitors.

In the 1970s in Europe and the USA, gin, vodka and so on attracted more consumers. This was the so-called 'White Revolution'. This was also experienced in Japan, and SHOCHU benefited. As a result of de-regulation and the liberalization of production, together with its association with health, SHOCHU attracted the interests of high-income earners. It began to be served in high-class Japanese restaurants and sold in stores specializing in alcoholic drinks. However, this change was by no means widespread. People were unable to taste the flavor of the SHOCHU and were also ignorant about appropriate drinking styles. It was in the period from 1975 to 1985 that the remarkable increase in SHOCHU production was seen. Since most SHOCHU makers were very small in 1960s, the industry was designated by the central government as an objective of 'the SM modernization project' (Takeuchi, 1985) and rationalization progressed, accompanied by subsidy. As a matter of course, in the course to promote the project, strengthening of R&D function by the maker's own efforts was progressed and it became the motive force in the next stage.

After the 1980s, the production of SHOCHU increased rapidly, stimulated by the OVOP project and rationalization took place quickly. SHOCHU makers began their own efforts to develop products, making the most of the sophisticated materials in each area and marketing products. Although SHOCHU production in Ohita prefecture has developed marvelously, it was a very new production area at that time. It was Kagoshima prefecture, located on the southern end of Kyushu, that initiated such a development project. The makers developed good-quality products by making use of sweet potatoes as the material and were eager to market their new product to the whole country. At the same time, distillers were willing to promote the ideal way to drink SHOCHU, to which special attention should be paid. Thanks to such a program, many Japanese learned how to drink SHOCHU. After that, makers in Miyazaki Prefecture started distilling of SHOCHU with buckwheat as the core material. Meanwhile, makers in Kumamoto Prefecture were ambitious enough to make rice-based SHOCHU. Thus various types of SHOCHU with new tastes were developed. Concurrently, the makers embarked on remodeling the bottles. They also invested in the broadcast of commercials on TV. As a result, consumption of type A SHOCHU was expanded among high-income groups in the metropolitan areas and other major cities. In this way, new customers of type A were not the people who habitually drank the inexpensive type B, but were the people converted from wine, whisky and SAKE consumption. SHOCHU not only came to be widely consumed, even in high-class Japanese restaurants, but was also sold in department stores.

An important issue for the SHOCHU industry was the tax problem. The consumption and import of Scotch whisky was declining and traders judged that all the trouble was caused by a tax regime that imposed very low tax on SHOCHU compared with that on whisky. Thus they filed an action through the British government and EU to the WTO, claiming that the low rate of tax on SHOCHU should be rectified. The United States and Canada, both of whom are exporters of bourbon to Japan, followed the action. The Japanese Government agreed in 1997 to make some changes and this naturally caused a great shock to the SHOCHU industry. The rate of tax on SHOCHU was raised 2.4 times and the rate on whisky was lowered by 60 per cent to make it equal to the rate for SHOCHU at present (Figure 9.3). Having succeeded in raising the tax on SHOCHU, Scotch traders naturally expected that the sale of whisky would increase. But, the tendency toward increased growth in SHOCHU production remained unchanged. On the other hand, demand for whisky continuously decreased. That is to say, the SHOCHU industry expanded by using the raising of the tax and price brought about by the pressure from Scotch traders as a springboard.

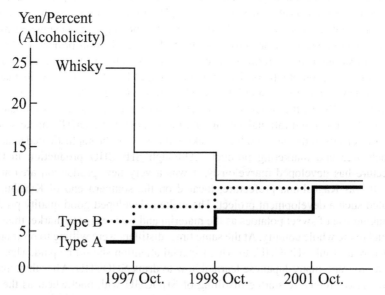

Figure 9.3   Change of liquor tax by product type in Japan

Contrary to all expectation, the increase in the price of SHOCHU accompanied by the additional tax generated a positive effect on consumption. That is to say, firstly, the erroneous image of SHOCHU as a cheap and low quality drink for the lower class was wiped out and it gained a good reputation among consumers as a high-class drink. This was linked to the rapid increase in consumption. Secondly, its good reputation was underpinned by the efforts to improve its taste or modifying the design of bottles. For example, the traditional porcelains of Kyushu were used as a design on some bottles. Some other bottles included designs made with high-quality colored glass. This resulted in an expansion of SHOCHU demand, ranging from major cities to places all over the country and it emerged as the preferred drink after beer. This trend occurred both at home and in restaurants. Every department store introduced a large sales display areas for SHOCHU. In this way, there were an increase in sales by raising the image of SHOCHU and consumers sought SHOCHU according to its brand. Thirdly, as a repercussion of the sale of Type A, demand for Type B rapidly increased as well. In particularly 'CHU-HAI' a cocktail of type B into which fruit juices such as plum, apple and so on was blended, spread explosively among younger consumers. CHU-HAI has now become popular with young people in bars and at parties as an inexpensive drink originating from Japan. It is a new creation of youth culture. Ironically, the lobby for Scotch whisky has played as an actor for the enhanced development of the SHOCHU industry, and for the formation of new consumption culture.

It is important to note that no culture of tasting whisky has intrinsically been developed in Japan. In homes as well, attrative bottles of Scotch are usually displayed as ornaments. At present, very few homes consume whisky as a daily drink. Water-diluted whisky is consumed to the greatest extent in various parties in Japan, and very few people wish to taste genuine whisky. That is to say, whisky has never found a wide consumer base as an alcoholic drink. People are keenly interested in the relationship between drinking and health and it has been verified in recent years that SHOCHU is an excellent drink for maintaining one's health. This is now widely known across the country and it has become a principal cause for the increase in consumption. It can be said that these matters also became a very important and favourable window for the promotion of the OVOP project in Kyusyu.

Formation and Development of SHOCHU Makers

Formation of SHOCHU Makers in Ohita (Kyusyu) Stimulated by the OVOP Project

In its long history, Kyushu island has been proud of preserving its tradition of SHOCHU production and maintaining its consumption culture. However Ohita, one of the local provinces in Kyushu not only failed to produce SHOCHU but also had been unsuccessful in promoting its consumption. In this prefecture, SHOCHU production started in the 1980s stimulated by the governor-guided OVOP project.

Since that time, Ohita Prefecture has developed into the largest production area for SHOCHU in Japan. Representative producers of SHOCHU in Ohita, include IICHIKO (Sanwa) and NIKAIDO. Characteristic of SHOCHU production in Ohita is the fact that SAKE makers had already been in existence as a pre-existing industry and made advance into the SHOCHU industry. For example, IICHIKO was established by merging four local SAKE makers in 1959. Faced with the difficulty caused by the stabilized demand for SAKE, the enterprise successfully produced SHOCHU by standardizing its production. Taking up wheat as a material in 1978, IICHIKO was successful in developing its business by engaging in research and development. The enterprise furthermore expanded its production by taking advantage of the 'One Village/One Product campaign'. The situation is similar for other makers. Moreover IICHIKO was able to capture the Tokyo area as its main market, making use of wholesaler networks for exclusive SAKE sales from the beginning. The enterprise was furthermore ambitious enough to devise a strategy for expansion of the market to the whole country.

In Ohita where people were mostly drinkers of SAKE and are ignorant of the tradition of drinking SHOCHU different with other areas in Kyusyu island. There have existed only a few very small scale SHOCHU makers who depended on a very local market. In such a situation, SAKE makers availed themselves of the OVOP project and embarked on the production of SHOCHU. Thus, in parallel with the successful formation of a new large-scale production area, a new drinking culture was created. The product development of these makers was based on the sensitivity of SAKE making. Meanwhile in the sphere of sales, the market was gradually expanded using the existing sales networks in SAKE. Every maker engages exclusively in production of their original local brands. Rationalization of production is being promoted and recent enhancement has successfully resulted in the acquisition of ISO14001 approved.

Expansion of the SHOCHU Makers

In other prefectures in Kyushu that were also willing to adopt the OVOP project following Ohita, emphasis was given to the development of the SHOCHU industry as one of the most important regional policies. That is to say in Kagoshima, Kumamoto (Hachikubo, 1996), and Miyazaki all of which played a role as a forerunner of the SHOCHU boom as well, continual growth of distillers is in progress and research and development functions have been strengthened. Thus, these have grown to be one of the JIBA (localized) industries (Ide and Takeuchi, 1980). Concurrently with the evolution of tastes accompanied with the improvement of brewing technology, further development is taking place in the design of the bottles, labels, and boxes. Demand has been stimulated. All of the makers are achieving growth thanks to the SHOCHU boom. The distillers that have now been converted into giant enterprises are aggressive enough to make vigorous advertisement promotions in the mass media. They are now making efforts to export the SHOCHU as a spirit to compete with Whisky in Europe, Asia and the USA. In the case of a maker in Ohita where SAKE is concurrently manufactured, they have also moved into the production of wine. Thus, they have diversified their operations as witnessed by visitor management focusing on distillery visits and restaurants. The business heightens consumption of agricultural products, and plays an important role in vitalizing rural villages because they generate employment. A serious environmental problem related to SHOCHU production is the large volume of waste from production. Almost all of its waste was once thrown into the sea but it is now reprocessed and reutilized as medicines and livestock feed. Thus, management of production and environment is being undertaken side by side.

Conclusions

The regional industrial policy of the central government to disperse industry from the metropolitan areas to the provinces has failed to attain great effect in industrial expansion and economic growth in the latter. In such a situation, the OVOP project started by a local government in Kyushu has been successful in developing industry by taking advantage of local conditions. The project, which allows local people to be proud of their success, not only promotes economic activity and traditional culture but also bringing about the creation of a new culture. Accordingly, this might be called an effective regional policy where priority has been given to the local communities. The SHOCHU industry made improvements, thanks to the OVOP project, and has extended its national and international markets. That is to say, the project created by local government to stimulate the economy has heightened the ambitions of local business leaders. Furthermore, through innovation in the products, marketing, and service, the demand for SHOCHU has been widened to create a new drinking culture throughout the country. It is an important issue to evaluate the locality and the role played by local governments. In this sense as well, the role played by the OVOP project is quite significant. As a direction for future promotion of the SHOCHU industry the formation of a wide-range of sales networks is necessary.

The OVOP project is at present of interests and being introduced as part of the regional policies of many countries such as China, the Philippines, Malaysia, Thailand, and the USA. The spirit of the 'one-village/one-product project' is first founded on the promotion of the industry upheld by the history, culture and attitude of the people of each local area; environmental symbiosis is increasingly the premise. In this way, it is one of the effective industrial policies achieved by unifying local and global systems. Although the project and the example of SHOCHU production is simple enough, it provides a model of regional promotion in many countries, especially for the renewal of rural communities.

References

Hachikubo, K. (1996), 'Formation of SHOCYU producing area and change of market in Kumamoto', *Hosei-Chiri*, 24, pp. 36–50(Japanese).

Hachikubo, K. and Oda, H. (2000), 'Development of the brewing industry in Japan-The Restructuring of traditional industry and its implication for sustainable development', Paper presented for the IGU Commission on the Dynamics of Industrial Spaces, Dong Gang, China.

Ide, S. and Takeuchi, A. (1980), 'Jiba sangyo-Localized industry', in A.J.G. (ed.), *Geography of Japan*, Tokyo, Teikoku-shoin.

Takeuchi, A. (1985), 'Policy for small industry in Japan', in Sitt, V. (ed.), *Strategies for Small-scale Industries in Asia*, London, Longman.

Takeuchi, A. (1994), 'Location dynamics of Japanese semi-conductor industry in the rapid technological innovation', *Geographical Review of Japan*, 66–2, pp. 91–104.

Takeuchi, A. (1995), 'Industrial transformation and environmental impacts in Japan', *Report of Researches, Nippon Institute of Technology*, 25–3/4, pp. 1–13.

Takeuchi, A. (1996), 'Regional development policy and Technopolis in Japan', *Report of Researches, Nippon Institute of Technology*, 26–3, pp. 1–14.

Takeuchi, A. and Mori, H. (2001), 'The sustainable renovation of the industrial complex in inner Tokyo: a core of Japanese machinery industry', *Geographical Review of Japan*, 74–1, pp. 33–46.

PART III
Internal Regulation and Policies for Service

Chapter 10

Reluctant State, Decentralized Markets, and Under Developed Communities: The Construction of the Futures Trading Industry in Taiwan

Pin-Hsien Chen and Jinn-yuh Hsu

The emergence of the financial industry has become a dominant feature in the capitalist economic system since the breakdown of the Breton Woods agreement in the 1970s. Innovative financial commodities, in combination with the new information and communication technologies, have been stirring up the market structure and dramatically reshaping financial landscapes. There are three penetrating and cross-enhancing processes characterizing the current global financial system: deregulation, technological innovation and globalization (Martin, 1999). Deregulation or liberalization of financial markets opens up new geographical markets and encourages the development of new financial products. Under such circumstances, some key financial transactions occurred in a few global cities, while others dispersed to lower tier financial centers connected by electronic networks. Thanks to the development of information and communication technologies (ICT), it was argued that the majority of transactions have become virtualized and even blind to location (O'Brien, 1992). Among the innovations in ICT, a wide array of new financial instruments has appeared on the scene that facilitated greater spreading of risk (Boden, 2000). The most important product innovation since the mid-1980s has been the phenomenal growth of the derivatives markets.[1] The goals of these markets were to enable investors to manage risk by offsetting the effects of volatility, and to allow them to gain profits by arbitraging the gap in information. However, as argued by Tickell (2000), derivatives not only allow institutions to offset risk but have also been implicated in a series of high profile losses which have undermined the financial viability of companies, banks and government entities (see also Pryke and

---

1 Derivatives are 'financial tools derived from other financial products, such as equities and currencies. The most common of these are futures, swaps, and options…The derivatives market aims to enable participants to manage their exposure to the risk of movements in interest rates, equities, and currencies' (Kelly, 1995).

Allen, 2000). So, how to survive and even leverage futures markets has become the key issue for financial institutions seeking to grow in the turbulent world market.

However, even as abstract as derivatives products are, they are still regulated and affected, to some degree, by the embedded institutions (Agnes, 2000). Just as there are different national socio-institutional variants of capitalism, so there are divergent modes of financial development in various national systems, even though the basic monetary systems have tended towards convergence in the era of the neo-liberalist hegemony. Of critical importance are differences in regulatory environments, which are under attack from the forces of deregulation but still affect financial activities and monetary networks in/across divergent national systems in so far as financial systems are regulatory spaces (Martin, 1994). But why has the state been willing to deregulate financial control?

This issue is particularly relevant in the case of East Asian developmental states such as Korea and Taiwan. A number of researchers, such as Wade (1990), Amsden (1989, 2001) and Evans (1995), claim that the miraculous development of East Asia's Newly Industrializing Countries (NIC) should be understood as a process in which the state plays a strategic role in taming domestic and international market forces and harnessing them for national ends. Fundamental to these miracles is an insistence on industrialization, rather than on following current comparative advantage. In other words, market allocation rationality is subjected to the priority of industrialization, which usually means employing new technologies to increase capital accumulation (Storper and Walker, 1989). Key to the process of industrialization is that the state lends directional thrust to the operations of the market mechanism.

Among the protagonists of developmental state theory, Wade (1990) carefully distinguishes state-leadership from market leadership in order to allow theorization of the complexity of state-market interactions. Based on an empirical study of Taiwan's industrialization process before 1980, Wade concentrates on the synergistic relationships between state activities and market allocation in each industrial sector. He provides a vivid example of how state intervention, under certain historical conditions, can help to kick start industrialization and create economic spaces within which the market mechanism can function. In other words, the state in the Asian NIC is not a passive administration that takes price signals for granted and does not take charge of a sound regulatory environment only. Rather, the developmental state is interventionalist and actually performs a very active role in targeting industrial development (Pempel, 1997).

However, the key to the developmental state leveraging its capability for realizing its goals is the financial pocket.[2] As the leading figure promoting this concept, Johnson (1987) argued that state control of finance was the most important, if not the defining, aspect of the developmental state, followed by other aspects such

---

2   As argued by Skocpol et al. (1985), a state's means of raising and deploying financial resources tell us more than could any other single factor about its existing capacities to create or strengthen state organizations, to co-opt political support, to subsidize economic enterprises and to fund social programs.

as labor relations, autonomy of the economic bureaucracy, and the combination of incentives and command structures. In fact, the states still performs the key role of setting the principal parameters of national monetary activities and the modes for integrating them with the international system (Thrift and Leyshon, 1997). They can exploit their administrative powers to discriminate between financial institution by limiting the range of financial products and markets in which different types of institutions may trade. Finance is the tie that binds the state to the industrialists in the developmental state. The state takes advantage of the monopoly of financial resources, and maneuvers to allocate financial credits to promote those industrial sectors which it targets for some reason. By doing so, the state can exert influence over the economy's investment pattern and guide sectoral mobility, since in such a financial structure firms rely on bank credits, which are under the control of the state, for raising finance beyond retained earnings and to respond quickly to the state's policy, as expressed through interest rate and other financial policies (Zysman, 1983). As a result, finance itself is conceived of as the effective conduit of industrial policy. The top concern for the developmental bureaucracy in relation to financial development is therefore the stability of the system; anything but the competitiveness of the financial sector. The students of the developmental state believe that the model can work well for the newly industrializing countries, enabling them to catch up and helping to propel miraculous growth, as demonstrated by the East Asian case (Wade, 1990).

In contrast, the phenomena of financial repression, or the intervention in financial markets by the developmental state, have been targeted as the key area for deregulation, or back to basics, by neoclassic economists (McKinnon, 1989).[3] The neo-liberalist rhetoric argues that 'external' intervention will distort the optimal operation of the market mechanism, and lead to financial dualism; the co-existence of the legal and the underground financial systems. Meanwhile, the argument goes, financial resources will be misallocated and rent seekers will emerge in the market (McKinnon, 1989). According to the neo-liberalist prescription, the government would be better keeping away from resource allocation and loosening financial control to allow prices (?) back to the right track.[4]

---

3   The state usually intervenes in four areas of financial activities to promote economic development: interest rate control, credit allocation control, financial institution regulation, and foreign exchange rate control (McKinnon, 1989).

4   The antithesis proposed by these two poles leads to divergent positions toward the explanation about the cause, as well as the antidote, of the East Asian financial crisis. For the neo-liberalists such as Greenspan (1998), Asia's sudden straits are symptoms of chronicle diseases of 'crony capitalism'. They accuse the 'picking up the winner' industrial policy of distorting investment capital, and centering on political patronage and close personal connections between powerful politicians, bankers, regulators and business people. As a result, poor quality regulation exists within many Asian financial institutions, and leads to moral hazard, public guarantee for nominally private business transactions. Thus, the cure for the 'over-politicized' disease is to leave the market alone. Again, a starkly divergent diagnosis and prescription is proposed by the believers of the developmental state. They argue that the

The debate raises an interesting issue: why would the developmental state engage in deregulation policy if the control of finance has been the lifeblood for it to survival and even prosperity? Why would states dismantle the institutions that gave them the power to exert control over financial resources, surrendering much of that power to the market? As illustrated above, two explanations stand out. The first privileges the role of market forces and the second underscores the importance of policy decisions being influenced by the advocacy of domestic business groups and international hegemonic institutions, particularly the US government. The former tends to explain financial liberalization and deregulation as a relentless transformative process driven by the development of new technologies, particularly new information and communication technologies, and more sophisticated financial instruments. As a result, the state cannot but accept the market rule and losing its control as a result. By contrast, the latter focuses on the politics of the alliance between local and extra-local economic forces that exerts an effect on policy making. It leads to the erosion of state autonomy and the decline of state capacity in the political struggle and consequently the state is forced to withdraw from the policy of financial repression.

These two perspectives are usually mobilized to explain the respective cases (Loriaux, 1997). Even though they represent contradictory viewpoints on the causes and effects of financial liberalization, they ironically share an unpredictable consensus that market forces and state bureaucracy could work well together provided they do not interfere with each other. However, it is widely agreed that developmental states have to embed themselves in the market in order to devise effective policies (Evans, 1995) and, at the same time, the de-regulated market needs re-regulation from the state to operate smoothly at different geographical scales: from local, regional to global levels (Weiss, 1998, Martin and Sunley, 1997). To set the tone as 'either or' for the two institutions creates a deadlock for further theoretical development (Boyer and Drache, 1996). In fact, the wisdom of recent institutionalism demonstrates that both of them will encounter the issues of information incompleteness, coordination and governance, and as a result usually fail to function well. Moreover, the institutions have their own ways of doing things and constitute barriers to radical change (Amin, 1999, Scott, 1995, Nelson and Winter, 1982).

In light of the institutional analysis, a number of intermediary institutions could allow a richer historical study of industrial development in a particular locality (Piore and Sabel, 1984, Saxenian, 1994). Among these meso-level institutions, the community is usually referred as the alternative, but complementary, mode

---

model of developmental state works well if without the conspiracy of American hegemonic neo-liberalism and its puppets, IMF (International Monetary Foundation) and World Bank (Wade and Veneroso, 1998, Panitch, 2000). Under the pressure of the triple unholy alliance, the exemplars of developmental state such as Korea are forced to withdraw their interventional policies, and start to deregulate their financial control. Consequently, the system becomes vulnerable to the attack of hot money, and runs into a catastrophe. In other words, it make nonsense to engage a deep restructuring of financial system because of a temporary liquidity crisis struck them as inappropriate, given the model has proven its manifold developmental advantage (Pempel, 1999).

of organization and mediates between the seemingly polarized position of state regulation and market rule.[5] As argued by Storper (2004), the particular local and national forms of social organization in which social actors' daily lives are embedded underpin the supposedly transparent and anonymous forces of markets and state in modern society. In turn, these two forces shape the particularity of the social organization and interaction within the community.

In the case of financial services, the formation of the community will be heavily affected by state action and market shaping, and vice versa. In a late-developing country such as Taiwan the financial market is strictly regulated at an early stage and has been forced to loose control gradually. What are the forces behind the liberalization process? Furthermore, what kinds of institutions will work with the state and the market to govern the new system? Moreover, the technology transfer literature has demonstrated the particularity of the latecomer context in industrial development (cf. Amsden, 1989) and this leads to the question as to whether the late development situation applies to the financial industry in the construction of market. If so, are there any differences for the development of the professional community in the local context? All of these questions will be explored in the research reported here which is based on a case study of the creation of Taiwan's futures trading sector.

In the next section, the methodology is explained, followed with a brief discussion of the relationship between the developmental state and financial repression before the liberalization of policy in the late 1980s. Then, we will detail the development of the futures market in Taiwan. A number of particularities will be raised to illustrate the role of the state in the shaping of the market and the associated professional community. After the historical review, the social networks and product innovation of the community will be discussed. It will focus on the three domains of social networks in the decoding of financial information in the context of Taiwan's futures market.

Methodology

This chapter is based on the findings of a research project that commenced in May, 2004. The project examined the social and professional networks of the futures trading industry in Taiwan. In the research process, interviews were conducted with the general managers of 28 futures trading companies selected from the membership list for the Taiwan Futures Company Association, the major industrial association. Besides, representatives of key government agencies, business organizations and other social groups in the futures industry, related financial sectors such as securities and banking were also consulted. At the start of the research, a preliminary working hypothesis was adopted about the relationship between state policy and the evolution

---

5   'Community' is widely referred as the networks of social interaction among social agents with specific identities or collective emotion (Bauman, 2001). They are bound by relations of common interest, purpose, or passion, and held together by routines and varying degrees of mutuality.

of the futures trading community in Taiwan's newly emerging market. This was used a framework for formulating a semi-structured set of interview questions. The interim framework allowed us to focus on the research topic, and at the same time, to expose us to feedback on our original hypotheses from the real world. As more information was gathered, we sharpened our analysis and raised more critical issues in the interviews. In the corporate interviews, in particular, we were cautious to double-check the results from each interview with cross-references. These interviews typically lasted at least one hour. After each interview, the researchers discussed the findings, encoded the information, and challenged each other's assessment of the topics discussed.

In addition to the in-depth interviews, government publications, business surveys, and journal reports have provided valuable input to this research. However, in order to avoid turning the research into a collection of business anecdotes, we have been particularly careful when attempting to draw conclusions from these reports. We have double-checked with the relevant people or agencies before making any final judgments.

From Financial Depression to Financial Deregulation?

As a developmental state, the then ruling party, the KMT party (Kuomintang), manipulated financial resources as the key tool to leverage strategic industries. The financial policy was very conservative under KMT rule, perhaps because of its experience of failure in Mainland China before it was defeated by the Communist Party in 1949. Because of attributing the failure partly to the financial disorder, the KMT party tightly controlled the development of financial activities after it retreated to Taiwan (Cheng, 1990). Three points, particularly relevant to the futures trading industry, characterized the financial repression in Taiwan before the liberalization policy of the late 1980s.

Firstly, financial activities have played an ambivalent role in economic development since the state initiated industrialization in 1960s. On the one hand, the state used its financial resources to target key industrial sectors in line with developmental state theory. Finance played the supporting role for national industrial policy. On the other hand, the state had to tame the financial beast by controlling its full fluidity in order to weaken its subverting potential. Private companies were not allowed to obtain foreign loans directly and had to check with the state-owned banks. The penetration of foreign capital was strictly limited to certain financial areas and the foreign exchange rate was set and manipulated by the government (Wang, 1996). Taiwan's economic growth has been financed almost exclusively by domestic savings.

Secondly, a banking system that controlled credit allocation was the pillar for Taiwan's financial system before late 1980s.[6] State-owned banks selected their

---

6   Zysman (1983) makes two main types of national financial systems. The one is a capital market dominated system, in which securities and bond markets are the major channel

clients not on the basis of level of interest rate they could levy, but on the basis of their own 'policy discretion'. More importantly, as securities and foreign exchange markets were strictly manipulated so as not to disturb financial stability, most of the funding for the private sector came from banking loans, rather than from the capital markets (Chu, 1999, Wu, 1993).

Finally, the weakness of the capital market led households to choose bank savings as the most convenient channel for investment, even though the official interest rate was undervalued (Wu, 1993). As the banking system could not absorb the amount of capital accumulated as a byproduct of the process of national economic growth, a number of illegal financial mechanisms began to emerge to 'solve' the problem of over-accumulation.[7]

However, things changed gradually following the liberalization policy of the late 1980s. On the one hand, economic restructuring started to push the majority of Taiwanese labor-intensive sectors, such as plastics, footwear, and textile industries, into seeking cheap labor. Information Technology (IT) industries became the propelling sector, and started to take advantage of the booming domestic savings in the capital market to rise the funds that they needed. Prices on the stock market skyrocketed and the economy overheated in the 1990s. On the other hand, the pressure from the US to open the market in financial areas forced Taiwan to change its policies on foreign exchange and regulation. In fact, the pressure for deregulation also came from key national business groups, which were excluded from setting up commercial banks, but which began to penetrate the financial forbidden garden in late 1980s (Lin, T., 2001).

In 1994 the state proposed a major project (Project of Asian Pacific Operation Center) to take on the challenges from both the domestic and international sides. A financial center plan was included in the Project. It aimed to loosen the regulation of the banking system to allow the emergence of big private banks as well promoting the operation of a 'sound' capital market to channel capital into profitable investments. In particular, the core objective for the financial center project was to transform the old banking-dominated system into one that was capital market dominated with a focus on the development of derivatives, foreign exchange, and stock markets (Yeh, 1997). Financial deregulation in the early 1990s led to a boom in the capital market

---

of capital raising. It's popular in the US and the UK. The other is a credit based system, in which the state or a small number of big banks control the interest rate and credit allocation, and other kinds of financial tools are difficult to be utilized for capital raising. It's prevalent in the East Asian developmental states such as Japan and Taiwan.

7   Most of the illegal investment companies collected capital from the community networks, with a name of *Mouse Association* to illustrate the prevalence of informal social networks. They offered a much higher interest rate (mostly double or even five times than those from the banks) to attract the household savings. They invested in the bubble economies such as land speculation. Once the economic condition encountered downturn, they went bankruptcy, and caused serious social problems (Lin, T., 2001). The underground futures trading companies, as shown below, is another case.

and around mid-1997 Taiwan's stock market ranked sixth in the world in average trading volume and fifteenth in terms of overall capitalization.

The change appears quite dramatic but why did the developmental state leave its major tool, credit allocation through the banking system, broken down without protection? In fact, in response to the neo-liberalism challenge, the state has chosen a sequence of financial liberalization steps that put priority on deregulating the domestic capital market over foreign participation. Despite the trend toward global financial integration, the state has still been cautious about the full internationalization of local currency, and controlled the volatility of cross-border movement of short-term capital (Chu, 1998). Even the QFIIs (Qualified Foreign Institutional Investors) were only allowed to enter the local capital market in an incremental rather than 'big bang' way. The state put a ceiling on the rate of stock held by the QFIIs in each local company's capital value at 5 per cent in 1991, rising to 15 per cent in 1998. Purchase of more than 50 per cent of any local company's value with local capital was not permitted until late 1999. In the adjustment process, small increments to the ceiling were usually increased by the state every two months. The process of financial deregulation in Taiwan was thus characterized by Chu (1998) as 'dragged and incremental'.

The reasons behind the conservative policy came from the contradiction between the need for financial stability *and* financial liberalization. The state tried to balance the two contradictory demands and created a catch-22 situation. Concerns for stability have become keen as Taiwan was diplomatically isolated from the international community and was in a situation of political and military confrontation with mainland China. The concerns caused the government to keep a wary eye on capital flows since:

> unlike Thailand, South Korea, Indonesia and the rest of the reckless tigers, Taiwan knew that it could not afford to go bust. As a diplomatic pariah, it was banished from virtually all important international organizations, including the IMF. If it got into trouble, it could not hope for a bail-out. Indeed, it worried that in the event of a domestic economic crisis the mainland China might just invade it' (*Economist*, Nov. 07, 1998).[8]

In other words, the state's longstanding policy guideline for financial management has been characterized by its overriding concern for financial stability since the KMT regime restored in Taiwan after 1949. However, the attitude towards the financial

---

8   Such cautious tones were echoed by the government even the political change after the presidential election in 2000, which ended the 50 years long of KMT rule in Taiwan. The new minister of finance, Mr. Hsu Chia-dong, claimed that 'in the antagonistic situation between mainland China and Taiwan, we could not but worrying the free flow of capital would hurt Taiwan's financial stability. If we were to let capital flow without ceiling, one the one hand, the mainland China would absorb most of capital outflow, and would keep Taiwanese capital hostage and threaten us. On the other hand, the capital from mainland China would enter Taiwan, and could disturb our financial order to match the military actions' (Lin, Y., 2001).

sector has shifted from seeing it as a tool for the industrialization goal under the thumb of the developmental state before the 1980s, to liberalizing it to allow growth in response to the neo-liberalism hegemony. The contradiction between the institutional path dependency for stability and the institutional change for growth constituted the core issue in the emergence of the capital markets. Without doubt, the dilemma also appeared in the decision to establish a futures trading industry, a high risk and abstract capital market.

The Rise of the Futures Trade Industry in Taiwan: Reluctant State and Decentralized Market

A Short History of Taiwan's Futures Market: Offshore First, Then Inshore

The burgeoning of Taiwan's futures market can be tracked to the 'underground' futures trades undertaken by a number of business groups from Hong Kong in the early 1970s. It was not legalized at that time, as the government viewed such activity as gambling and did not want it to disturb financial stability. The accumulated household savings aggravated this money game and stimulated the growth of an underground market in the late 1980s.[9] The underground dealing companies were generally criticized as bucket shops, not sending clients' orders to foreign exchanges, parceling clients' money, and running away when they lost clients' money (Li, 1982).[10] The Taiwan government had twice (1983 and 1996) raided these futures companies but failed to wipe them out. Even though the legal FCM (Futures Commission Merchants) or IB (Introducing Broker) are widespread today, the underground futures companies have never perished (as reported in some newspapers, see Yang 2003). There were over 350 underground traders and a trading volume of more than NT$ 40 billion per day at the peak (Lin, 1988). It was not until the early 1990s that the government decided to legalize the futures trading industry (FTI).

The decision in the late 1980s to introduce legislation was made at a time when staple commodities price instability caused by international currency fluctuation was an important problem; futures trading was suggested by several key commodity exporters as a way of hedging the risk. Moreover, from then on, the Taiwan

---

9   Chu and Lee (1998) argued that a series of ominous economic signs, such as the mushrooming of underground financial institutions, bubbles in real estate and stock markets, and rapid deterioration of private sector investment, compelled the government to take decisive measures to overhaul the financial architecture in late 1980s.

10   According to Mr. L, the president of company H, a futures house with 120 employees, who has worked in the underground shops for more than five years, 'the underground dealers cheated their customers in most cases. When the customers deposited a margin, let's say NT$ one million, the dealers took more than half as their profit. If the customers lose, the money would stay in the dealers' pockets. Even if the customers won, the dealers would close the shop and run away to wait time to open another new shop. It caused serious issues in the public order' (interview with Mr. L, July 9, 2004).

economy has faced the pressure of industrial restructuring, and service industries, including financial services, have become critical for economic development. In addition, the trading volume on the stock market has continuously expanded and sometimes fluctuated fiercely, which also implies an increasing need for investment diversification and hedging. The relevant legislation finally passed into law in 1993 and the first futures brokerage company, Grand Cathay Futures Corporation, was established in 1994.

However, the legacy of the underground companies for the futures industry was unexpectedly significant in two perspectives. On the one hand, it served as a training school for many of the key managers of the 'legal' trading companies. Most of the 'underground' ex-employees accumulated their brokerage experiences and mastered the skills, as well as languages, of operation at that time. Once the industry was institutionalized, they were recruited as the foundation employees, and were soon promoted to the manager ranks. In a sense, the underground companies trained the generals and paved the way for the later legal 'troops'.

On the other hand, the fraud practices endemic during the underground period left a scar in the minds of the general public, and impinged on the government's policy making. In fact, underground traders still exist in Taiwan in order to take advantage of low brokerage fees and narrower margins than those in the legitimate businesses.[11] To avoid fueling the money game, the government adopts a gradualist approach to financial deregulation. Rather than pushing the market growth hard, the government loosened financial control step by step. It basically reacted to the issue of underground futures as a social problem, without a clear intention to take the futures market as a key element in the construction of a sound financial system. Under such circumstance, the futures markets were legalized in two steps: first, offshore futures contracts trading, such as those launched by the Chicago and Singapore markets, were permitted and this bypassed the 'Offshore Futures Transaction Law' of 1994. After three years, an inshore trading market, the Taiwan Futures Exchange Corp. (TAIFEX) was established in 1997, and eventually in July 1998 the first local futures contract designed by TAIFEX was launched.

The legislative process was not without controversy. Establishing the FTI was taken as part of the project for setting up the regional financial center in Taipei. A hot debate about the sequence of opening either offshore or domestic markets first occurred in the legislation process. The government finally opened the offshore transactions first because the underground traders had engaged in such business for a long time and it would be easier to legalize such transactions. If the policy was to prioritize the domestic market, which would be beneficial for the long-term objective for a sound financial system, it would take longer to pass the necessary

---

11 According to the law, each trader has to provide an initial margin for transaction to guarantee the later payment. In most cases, it only half or even none of the initial margin is required in the underground practice, which is usually based on mutual trust. It is obvious that the transaction cost is much lower for the underground brokers. As a result, it is expected that the underground brokers over-number their legal counterparts today.

legislation and might trigger speculation (Zan, 1993). Even in the permission for the offshore transactions, the capital requirement for establishing a brokerage house is NT$ 200 million (equivalent to US $ 7 million), quite high by comparison with their counterparts in USA, which requires only US$ 250 thousand.[12] This reflects again the logic behind the government's reluctant internationalization policy: to control the volatility of cross-border movement of short-term capital and hinder the full penetration of local financial activities by foreign capital.

However, these initiatives were challenged by the lawmakers who represented a number of key business groups. Among the disputes, the issue of whether to allow the same business group to engage in both stock dealing and futures dealing was raised. According to the government's plan, it was not permitted to merge together to create securities houses that would be beyond governmental control. But, as the securities business groups gathered together to fight, the government made slight adjustments to its initiatives that allowed securities houses to spin off another formally independent futures shops. 'By doing so, a firewall could be raised to prevent illegal dealings to occur for the financial business' argued by Mr. Lin Zong-Yong, the then vice-chair of the committee of securities management, who was in charge of the establishment of the FTI. The government was indeed cautious about the opening of seemingly speculative business and tried setting up barriers to the merger of financial businesses to avoid the prospect of disturbing the financial order.

In addition to the restrictions on market area and starting capital, the law also regulated closely the kinds of transaction and contracting procedure. Only those complying with the law will be allowed to engage, otherwise, a penalty will be imposed. In the words of Mr. W, the chair of the futures association:

> the aim of the government in the futures trade law is to prevent abuse, rather than to promote the industry. The bureaucrats do not take the industry as a necessary element for the capital market to grow healthy, but as a trap of speculation. Even they are forced to open market, they hesitate to comply with the liberalization doctrine'.[13]

Irrespective of the extent to which the policy objectives materialized, two positive effects were certain for the development of futures trading market in Taiwan. One was the change in the public image of futures products and activities. At least, it was no longer exclusively illegal and could be a normal channel of investment. More than 22 futures trading companies were established in March 1995, one year after the passing of the law. The other was the creation of a professional association, the Taipei Futures Trader Industrial Association, for the futures industry. In order to control the development of the sector it was stipulated that all of the futures houses had to join the association. It represents their collective interest and engages in the promotion of the industry.

---

12 In fact, the high capital requirement of the legal trader hinders the legalization of most underground traders. As a result, it is reported that most of the underground traders remain untouched after the law passed, and even keep growing today (Gao, 1995).

13 Interview with Mr. W, July 21, 2004.

Nevertheless, the most significant development in the futures market has been the establishment and growth of the local exchange market. After permitting foreign futures transactions, the government started to draft the law to set up a local trade market. Meanwhile, two events pushed the government to speed up the legislation process. Firstly, a Taiwan Stock Index Futures planned to be launched by the CME and SIMEX in late 1996 caused concerns among the state if Taiwan would lose the supposedly advantageous local merchandize.[14] Secondly, an idea for a 'Pacific Asian Regional Operation Center' was proposed by the state to situate Taiwan in the global economic system. It argued that in order to attract foreign investment and to push the domestic financial market and institutions to be international, the government should head toward a journey of deregulation and liberation. A local futures market was taken as one of the symbols of a competitive financial system. Besides, the newly founded Futures Traders Association lobbied heavily with the opposition party legislators and forced the government hand to prompt the legislation process. Under the circumstance, the Futures Trade Law, which mainly covered the establishment of a local futures exchange market, was passed in 1998. Under the law, a locally launched index, the Taiwan Securities Exchange Capitalization Weighted Index (TAIEX) became a major target in the local market. A series of measures in Taiwan's futures market are shown in Table 10.1.

After the opening of the local market, the scale of Taiwan futures market has kept growing. Figure 10.1 shows the total trading volume of the Taiwan futures market from August 1998 to March 2004 and the following Table 10.2 shows the trading volume of each contract in TAIFEX. Figure 10.1 shows how the trading volume at the Taiwan Futures Exchange has skyrocketed, as the volume in the TAIEX options contracts repeatedly jumped higher. In addition, the TAIEX futures contract also fared well. As a result of such growth, the Taiwan Futures Exchange joined the top 20 derivatives exchanges by volume for the first time in the first quarter of 2004 (DeGrandis, 2004).

Such stock-based products reflect one of the major characteristics of Taiwan's economy: the turnover and volatility of the local stock market is extremely high in comparison with other countries. Since its takeoff in 1994, Taiwan's over-the-counter market has rapidly expanded into one of the most highly capitalized OTC markets in the world. An anticipation of the vigorous stock market influenced the planning and launching the sequence of derivatives; the government wanted to avoid fueling an overheating stock market by promoting the futures market too hard (Zou, 2002). The

14   In early 1996, SIMEX in Singapore has contacted with the Securities Management Commission (SMC) for their plan to launch the Taiwan Stock Index Futures in their market. The same year later, CME in Chicago also showed their intention to do the same commodity. As the law for the local transaction was yet passed, the government did not approve the actions taken by the SIMEX and CME. An urgent meeting was called by the SMC to coordinate and dissuade these 'unfriendly' actions. At the same time, the government decided to accelerate the passing of the law to legalize the local transactions, 'so that Taiwan could remain competitive in the establishment of the regional financial center', according to Lu Dong Ying, then the chair of SMC (Li, 1997).

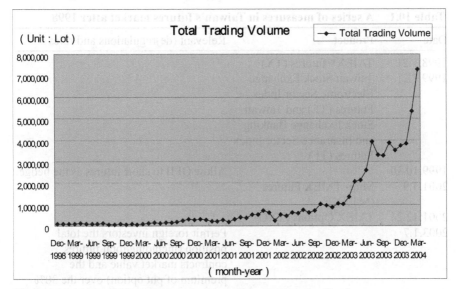

Figure 10.1   Total of trading volume at Taiwan Futures Exchange

Securities and Futures Commission was set up to manage the development of these two industries in 1997. The Commission vigilantly launched new commodities and little by little loosened the controls on the penetration of foreign capital into the local market. Mr. L1, the president of S Futures, the third largest house in Taiwan and a senior veteran in the industry complained that:

> the government never gave the futures industry an equal footing with the securities, and neither took these two industries as symbiotic. The key concern for the Securities and Futures Commission has always targeted on the development of stock market, and the futures market next. Every time while the Commission decided to launch new commodities in the latter, the first question would be if these would disturb the former.[15]

Market Facts

In line with the wisdom of institutionalism, the market was embedded in social and political structures. State policies would shape the market behavior in some ways. The reluctant liberalization enforced by the state resulted in the weakness of the financial service industries in Taiwan. As shown above, the state set up strict rules on the allowance of investment in the futures market by the institutional investors, so that the state could tame the organized investors safely. In combination with plenty of household savings without appropriate investment outlets, futures and options trading, similar to the structure of the stock market, was driven primarily by natural

---

15   Interview with Mr. L1, July 9, 2004.

**Table 10.1	A series of measures in Taiwan's futures market after 1998**

Date	Product	Relevant (de)regulations and rules
1998.7.21	TAIEX* futures (TX)	
1999.7.21	Taiwan Stock Exchange Electronic Sector Index Futures (TE) and Taiwan Stock Exchange Banking and Insurance Sector Index Futures (TF)	
1999.10.30		Allow QFII to short futures as the hedge.
2001.4.9	Mini- TAIEX Futures (MTX)	
2001.12.24	TAIEX Options (TXO)	
2003.1.7		Permit foreign investors the total position (including total futures contracts market value and the premium of put option) over the 30% of the purchased stock.
2003.1.20	Equity Options (STO)	
2003.6.30	Taiwan 50 futures (T5F)	
2004.1.2	10-year Government Bond Futures (GBF)	
2004.5.21		Allow foreign investors to long futures as the hedge to the position of securities to be purchased.
2004.5.31	30-Day Commercial Paper Interest Rate Futures (CPF)	
2004.6.15		Allow foreign investors to long futures as the hedge, regardless of whether they have already invested in spot market.
2004.6.30		Stipulate foreign investor for the long and short position separately. (The decree induced serious reaction from futures association and those institutional investors influenced.)
2004.7.15		In the newsletter issued by the Securities and Futures Bureau (SFB), the SFB approved of some opinions of the futures association, and hence promised to use net position to regulate foreign investors. But the details of calculation were still left to consult.

*'TAIEX': Taiwan Stock Exchange Capitalization Weighted Stock Index.
Source: http://www.sfb.gov.tw/ensfcindex.htm.

Table 10.2   Trading volume of each contract in TAIFEX

Futures Contracts	Total Trading Volume(Lot) 2004, Jan-Mar.	2003	2002	2001	2000	1999
TAIEX Futures (TX)		6,514,691	4,132,040	2,844,709	1,339,908	971,578
Taiwan Stock Exchange Electronic Sector Index Futures (TE)		990,752	834,920	684,862	409,706	87,156
Taiwan Stock Exchange Banking and Insurance Sector Index Futures (TF)		1,126,895	366,790	389,538	177,175	18,938
Mini-TAIEX Futures (MTX)		1,316,712	1,044,058	427,144	NA	NA
Taiwan 50 Futures (T5F)		4,068	NA	NA	NA	NA
TAIEX Options (TXO)		21,720,083	1,566,446	5,137	NA	NA
Equity Options (STO)		201,733	NA	NA	NA	NA
10G bonds		NA	NA	NA	NA	NA
Total	16,433,992	31,874,934	7,944,254	4,351,390	1,926,789	1,077,672

Source: TAIFEX, http://www.taifex.com.tw/taifex0302.asp

*NA:No data, because which contract was not launched yet in that year.

**The underlying index of both TX and MTX is Taiwan Stock Exchange Capitalization Weighted Stock Index (TAIEX). The TX is 4 times the contract size of the MTX.

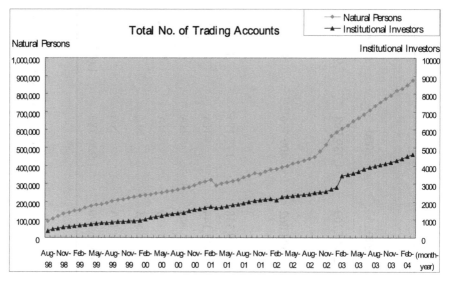

Figure 10.2 Trading account number of natural persons and institutional investors

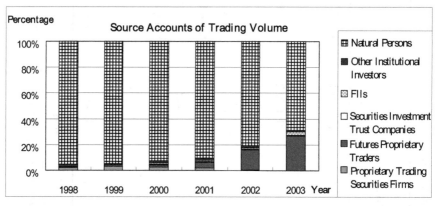

Figure 10.3 Source account of trading

persons (retail investors). In total trading volume, retail investors always accounted for over 50 per cent (Figures 10.2 and 10.3).

A market dominated by retail investors influenced heavily the activities and strategies of the futures dealers. Most of the individual investors carried less analytic capability and decision-making power in the market, and usually relied overwhelmingly on the judgments of the experts. As argued by Leyshon, Thrift and Pratt (1998), the perennial problem which faces all producers of financial services is information asymmetry, that is, providers and consumers of financial products have unequal amounts of information (and knowledge) about whether or not customers have the wherewithal to make them 'capable' purchasers. Natural persons are usually conceived as seriously information-handicapped investors, by comparison with institutional investors who usually employ researcher and analyst teams to help decision-making in an increasingly complex financial world. In addition, the natural persons are also quite sensitive to price, they care most about the cost of brokerage fees and transaction tax. Price competition rules over the futures market in Taiwan.[16] In other words, the major group of customers in the market is typically short of expertise/knowledge and susceptible to making decisions on the basis mainly of price. Therefore, two effects are produced in this market dominated by retail investors. Firstly, it leads the futures traders to exploit the ICT (information and communication technology) which is often claimed to render the financial texts readable and empower individual investors in decision making, to cost down their business operation. As a result, the digitalization of financial services is in vogue in Taiwan. Designing a humanlike trade platform becomes one of the major advantages in Taiwan's futures market. Secondly, even though ICT is widely utilized in the market, 'traditional' interpersonal relations and networks still matter in this natural person dominated market. The trust engendered in the process of other financial transactions, particularly securities and banking, can be helpful for the futures traders to expand their customer bases.[17] Consequently, the IBs (introducing brokers) who work in the branches of securities houses but are engaged in futures dealings for their frequent customers still play a key, if declining role, in the operation of Taiwan's futures industry.

The other characteristic of Taiwan's futures market is that foreign capital plays a relatively insignificant role (Figure 10.3). The government still hesitates to remove the restrictions on the entry of foreign capital in the decision making process. It adopted a series of gradual and adjustable measures on the deregulation of foreign penetration (see Table 10.1). The attitude of xenophobia illustrates the concern of the government to buffer the local financial market from the incursion of skillful 'players' who might jeopardize the stability of the stock and futures markets. Not until very recently did the Taiwan government approve of the newly amended regulations on foreign investors in futures trading.[18]

---

16 Almost of all general managers complained the deteriorating price war in the business from our interviews. For example, the president of P Futures, the largest futures house, Mr. H described the severity of the price war, 'We introduced a humane interface technique which rendered the transaction easy from Korea to the market and set the brokerage fee at NT$ 100 per each transaction six months ago. Two months later, another company transferred the similar technology, and reduced the price to NT$ 80 per unit. This move forced me to follow suit and reduced the fee to NT$ 69. But, taking account of the cost of the technology transfer, the reasonable price should be around NT$70. The cut-throat competition would lead to an internecine situation' (interview with Mr. H, July 23, 2004).

17 Most of Taiwan's futures trading companies belong to the financial holding groups after 2003. Therefore, the former can get access to the customer bases of the latter which usually include banking and securities services.

18 Take the disputing order issued on June 30 2004 for example. The Premier Yu announced 4 liberalization measures on May 2004 to 'accelerate Taiwan's capital market

The ambivalent policies on foreign capital dissuade, to some extent, foreign investors from embracing Taiwan's market. Even though foreign capital takes a great deal of interest in the booming market, few of them set up offices in Taiwan. Most of the transactions are conducted from Hong Kong, with ordering contracts made through the channels of some specific traders, such as Capital and Grand Cathay. It is generally conceived that foreign capital, based on its long term accumulated experience, possess better trading skills than the local ones. The expatriate communities formed by the managers and professionals employed in the foreign companies could constitute a key channel of technology transfer to the host society in the financial centers, as shown by Beaverstock and Bostock (2000). As most foreign dealings are conducted 'at a distance', the story basically does not apply to the case of Taipei, with the exception that foreign capital has to dispatch its traders to Taipei for the 'market feel'. Nevertheless, the skilled transient international migrants still only constitutes a small group in the futures trading communities in Taiwan.

The Formation of the Under-developing Community of Professionals

Another key feature shaped by state policy was the communities of the futures dealers. So far, three major groups have been identified in the managing cohorts in Taiwan's futures trading industry. The first group is composed of those managers who worked in the underground shops at the early stage and were recruited to the newly established futures houses. Most of them accumulated the trading skills from nothing via the bucket shops; few have degrees in the relevant disciplines. Learning by doing was the key channel of mastering trading skills.

> In the beginning, I, as a graduate from agriculture studies, knew nothing about futures. As an underground company, it could not offer the formal course to train new employees. My then boss, who was from Hong Kong and has been in the business for more than 10 years,

---

internationalization, and to render Taiwan's market development more in line with advanced securities markets around the world' (SFC, 2004). However, the practical orders and actions showed the ambivalent attitudes and being unfamiliar with practices of trading/market. The origin was that on June 15, in furtherance of the Yu's speaking, the SFC amended relevant articles to relax varied limitations on foreign investors, which allowed foreign institutional investors to long futures as the hedge to the position of securities to be purchased, regardless of whether they have already invested in spot market. It was expected to increase the flexibility of foreign fund's investment strategy and operation. But on June 30, following the afore-mentioned relaxation, an auxiliary order required the long and short position counting respectively. Moreover, the total amount of long position, which meant futures contracts market value plus options premium, can't exceed the cash and approximate cash position. In addition, the short position can't be over the total market value of securities possessed (i.e. stock and bonds). Such order upset local FCM and foreign institutional investors a lot, not only because it separately counts long and short position, but also the high requirement for buying hedging. The futures association and some investment institutions (including Merrill Lynch and others) immediately tried to exchange views with officials.

took me directly to the trading place just after a short course. I spent time accumulating trading experiences and interacted closely with my customers and supervisor. It took me two years to master the variety of skills in handling different positions (Mr. L, the president of H Futures, recalled his early experience).[19]

As first movers, they were mostly promoted to high ranking positions in the newly legalized futures trading companies.

The second group came from those who participated in other financial service industries, particularly securities, before transferring to the futures sector. These experienced financial operators usually took advantage of the combination of varieties of financial tools in running the futures business. Most of them were employed by the newly established financial holding companies, and could utilize their social relations to persuade customers from other financial fields to join the futures market. They built up their competitive edge in the business by exploiting their widespread social networks within the futures industry and extending to other financial business at the entry stage. Accumulating knowledge about the stock market, they were usually quick to get particularly familiar with the trading skills in option futures. Based on their existing skills, they had to be sensitive about the difference between these two seemingly overlapping businesses. As stressed by Mr. C, the president of W Futures:

> The requirement of professionalism is higher in the futures industry than that in securities. While I was in the latter, what I needed to do is to establish relations with my customers, but in the futures business, I have to do research in trade model building and strategy setting, otherwise, I will be eliminated through competition.[20]

Besides the top managers, the makeup of the operating group was mainly the salespersons in the front counter, or the IBs (introducing brokers), in the securities houses. The securities sector has been better developed in Taiwan, and penetrated each household economy. It served as the capillary attraction of capital flow, and constituted one of the major channels for managing household finance. Most of the key financial groups had securities branches over the island. For the futures sector in the same group, these amounted to the could-not-be-better sales points. Take the case of company F, the currently number four market sharer, as an example. The F Securities possessed 64 branches, in contrast to only one for the F Futures. Thus, the latter provided a series of on-the-job training, and incentive mechanisms to mobilize the salespersons of the former to expand its market share. According to Mr. J, the president of F Futures, 'The interpersonal trust fabricated between the IBs and their customers from the securities business could be transferred to the futures business'.[21]

---

19 Interview with Mr. L, July 9, 2004.
20 Interview with Mr. C, July 14, 2004.
21 Interview with Mr. J, July 15, 2004.

The third group, which is relatively small, comprises the ex-employees of the foreign companies in Taiwan or abroad. Among them, Chair W of the futures Industry Association and Mr. L1 of S Futures are exemplars. Mr. L1 worked in Merrill Lynch (Taiwan) before hopping to local companies, and Mr. W was a long-time futures brokerage manager in Chicago before being recruited back to run the local company, and become the Chair of the Futures Association. Since the activities of foreign capital were constrained, these veterans of foreign companies were the key channels for transferring new products and operation techniques, such as hedging skills, back to Taiwan.

The institutionalization of the communities was achieved by the formation of the Taiwan Futures Company Association. As the membership was obligatory by the law, the association became the key agent of collective interest and identity. Through the participation of the social and professional activities sponsored by the association, these three groups of managers maintained and enhanced dense social networks. More important, their monthly meetings provided a forum for the members to settle disputes and to reach consensus. The association could play the role of collective agent to lobby for favorable policies in the legislation process (Liu 1996).

Overall, the communities of professionals in Taiwan's futures industry were dominated by the locally trained groups spun off from other financial businesses, mainly securities, with a few returnees equipped with foreign experiences and skills. As most of the transactions were based on cost consideration, the expertise for the futures dealers was to master the borrowed technique from other countries such as Korea, and to exploit the advantage of networking extending from other financial business. Under the circumstance, it was the extent of social networks, rather than the sophistication of trading skill that was critical in the business. In other words, the reluctant state policies led to a decentralized market structure which further shaped an under-developed professional community in Taiwan's futures trading industry.

Conclusion

While financial liberalization was praised by neo-liberalists and criticized by political economists, its process has not been thoroughly examined in most studies. The establishment of Taiwan's futures market provides a case example to show how a developmental state adopted policies to meet the liberalization challenge, and at the same time, maneuvered to maintain the stability. Against the neo-liberalist argument that liberalization policy demonstrated the death of the state in managing financial flow, Taiwan's case suggests that the state can contrive a sequence of liberalization to balance the deregulation mandate and financial order. The case also refutes the discourse of the developmental state which identified the bureaucratic plan as the key driver behind financial transformation, since dragged and reactive responses characterized the main theme of state policy. In the process of forced liberalization, a market dominated by natural persons and little foreign capital participation was created. The market was characterized as decentralized and price sensitive to the service charge, and at the same time, required the brokers to stitch the knowledge gap for individual investors, rather than the institutional ones, in the complex financial transactions. Under such circumstance, it was the extent of social networks, rather than the sophistication of trading skill that was critical in the business. In other words, the reluctant state policies led to a decentralized market structure, which further shaped an under-developed professional community in Taiwan's futures trading industry.

As Storper (1997) argued, institutions consist of a persistent and connected set of rules, formal and informal, that prescribe behavioral roles, constrain activities and shape expectations. Institutions provide a cognitive structure to legitimize collective action, and tend to evolve incrementally in a self-reproducing and continuity-preserving way (North, 1990). At the same time, institutions are the products of historically-situated interactions, conflicts, and negotiations among different socio-economic actors and groups (Martin, 2000). Thus, institutional breakthrough could occur in the course of political struggle among the asymmetrically powerful agents.

The dialectics of path dependency and path breakthrough can shed light on the disentangling of the deadlock of state-market antitheses. Following the theoretical thread, the liberalization of the financial system in the developmental state can be understood as a process of the state's will to adjust in the new historical conjuncture set by the neo-liberalist hegemony, and simultaneously, it wants to keep policy priority over the control of the financial resources to avoid the damage drawn from the abrupt de-regulation process. It's a process of institutional learning, as well as institutional resistance.

Moreover, institutional evolution unfolds in different ways at various geographical scales, and constitutes the pillar of the particularity of each locality (MacLeod, 2001). Martin (2000) added that the evolution of particular institutional architectures would give rise to distinctive local, regional and nationally instituted 'configurations of capitalism'.[22] Hayter (2004), furthermore, argued that path dependent sectoral, regional or corporate trajectories were neither predetermined nor 'random walks'. Institutional evolution matters in different ways in different places. To claim financial liberalization as universalism is, in a sense, misplaced. In contrast, identifying and scrutinizing the social agents that interact with each other and their situated institutional endowments in the process of the financial deregulation in specific cases could be stimulating in the study of financial geography (Clark, Tracey and Smith 2002).

References

Agnes, P. (2000), 'The 'end of geography' in financial services? Local embeddedness and territorialization in the interest rate swaps industry', *Economic Geography* 76(4): 347–366.

_____

22 Boyer (2000) made a similar argument on different modes of financial regulation in the different capitalist states. Also see Albert (1993) on the distinct financial systems in the Agro-Saxon model and Rhine model.

Albert, M. (1993), *Capitalism vs. Capitalism: How America's Obsession with Individual Achievement and Short-term Profit has Led it to the Brink of Collapse*, New York: Four Walls Eight Windows.

Amin, A. (1999), 'An institutionalist perspective on regional economic development', *International Journal of Urban and Regional Research* 23(2): 365–378.

Amsden, A. (1989), *Asia's Next Giant: South Korea and Late Industrialization*, New York: Oxford University Press.

Amsden, A. (2001), *The Rise of the Rest: Challenges to the West from Late-industrializing Economies*, Oxford: Oxford University Press.

Bauman, Z. (2001), *Community: Seeking Safety in an Insecure World*, Cambridge: Polity Press.

Beaverstock, J. and Bostock, R. (2000), 'Expatriate Communities in Asia-Pacific Financial Centres: The Case of Singapore', *GaWC Research Bulletin 27*.

Boden, D. (2000), 'Worlds in action: information, instantaneity and global futures trading', in B. Adam, U. Beck and J. Van Loon (eds), *The Risk Society and Beyond: Critical Issues for Social Theory.* London: Sage.

Boyer, R. and Drache, D. (1996), *States against Markets: The Limits of Globalization*, New York: Routledge.

Boyer, R. (2000), 'The political in the era of globalization and finance: focus on some regulation school research', *International Journal of Urban and Regional Research* 24(2): 274–322.

Cheng, Tun-jen (1990), 'Political regimes and development strategies: South Korea and Taiwan', in G. Gereffi and D. Wyman (eds), *Manufacturing Miracles*, Princeton: Princeton University Press.

Chu, Yun-han (1999), 'Surviving the East Asian financial storm: the political foundation of Taiwan's economic resilience', in T. Pempel (ed.), *The Politics of the Asian Economic Crisis*, Ithaca: Cornell University Press.

Chu, Yun-peng (1998), 'Review on the financial liberalization policies after 1980s: delay, leapfrogging, and evolution of learning mechanism', in the *Symposium on Taiwan's Economic Development after 1980s*, Taipei: Chung-hua Institute for Economic Research (in Chinese).

Chu, Yun-peng and Lee, Tung-hao (1998), 'From bubbles to new rounds of Asian monetary cooperation – with reference to the Taiwanese', paper presented at a conference on contemporary Taiwan, Sponsored by the National Policy Research, Taiwan.

Clark, G., Tracy, P. and Smith, L. (2002), 'Rethinking comparative studies: an agent-centered perspective', *Global Networks* 2(4): 263–284.

DeGrandis, M. (2004), 'Trading volume: U.S. futures trading soars', *Futures Industry Magazine*, July/August 2004: 1–3.

Evans, P. (1995), *Embedded Autonomy: States and Industrial Transformation*, Princeton: Princeton University Press.

Gao, Zhen-huang (1995), 'Taiwan will be excluded from the internationalization stream, if not to legalize the derivatives', *China Times*, March 13, 1995 (in Chinese).

Greenspan, A. (1998), 'Statement before the Committee on Banking and Financial Services, January 30, 1998 U.S. House of Representatives', *Federal Reserve Bulletin* 84(3): 186.

Hayter, R. (2004), 'Economic geography as dissenting institutionalism: the embeddedness, evolution and differentiation of regions', *Geografiska Annaler B* 86(2): 95-115.

Johnson, C. (1987), 'Political institutions and economic performance: the government-business relationship in Japan, South Korea, and Taiwan', in F. Deyo (ed.), *The Political Economy of the New Asian Industrialism*, Ithaca: Cornell University Press.

Kelly, R. (1995), 'Derivatives – a growing threat to the international financial system', in J. Michie and Grieve Smith (eds), *Managing the Global Economy*, Oxford: Oxford University Press.

Leyshon, A., Thrift, N. and Pratt, J. (1998), 'Reading financial services: texts, consumers, and financial literacy', *Environment and Planning D* 16: 29–55.

Li, Dong-zhu (1997), '16 futures houses call on to open the foreign trading', *Commercial Times*, March 5, 1997 (in Chinese).

Li, Sir (1982), 'The underground economy: the mouse that cuts the lifeblood of the Economy', *Common Wealth* 18: 11–20 (in Chinese).

Lin, Ting-yau (2001), 'New Cell, Old Gene: The Dragged Institutional Change of Financial system in Taiwan', Master Thesis. Department of Sociology, Tunghai University (in Chinese).

Lin, Yu-ling (2001), 'An interview with the minister of finance, Mr. Hsu Jia-Dong' in *China Times*, August 3, 2001.

Lin, Zhong-xong (1988), *The Forty Years Experience of Taiwan's Economic Development*, Taipei: Independence News Press (in Chinese).

Liu, Shen-fen (1996), 'The legalization of the futures trading service', *United News*, May 10, 1996 (in Chinese).

Loriaux, M. (1997), 'The end of credit activism in interventionalist states', in M. Loriaux (ed.) *Capital Ungoverned: Liberalizing Finance in Interventionalist States*, Ithaca: Cornell University Press.

MacLeod, G.. (2001), 'Beyond soft institutionalism: accumulation, regulation, and their geographical fixes', *Environment and Planning A* 33(7): 1145–1167.

Martin, R. and Sunley, P. (1997), 'The post-Keynesian state and the space economy', in R. Lee and J. Wills (eds), *Geographies of Economies*, London: Arnold.

Martin, R. (1994), 'Stateless monies, global financial integration and national economic autonomy: the end of geography?', in S. Corbridge, N. Thrift and R. Martin (eds), *Money, Power and Space*, Oxford: Basil Blackwell.

Martin, R. (1999), 'The new economic geography of money', in R. Martin (ed.), *Money and the Space Economy*, Chichester: John Wiley & Sons.

Martin, R. (2000), 'Institutional approaches in economic geography', in T. Barnes and E. Sheppard (eds), *Companion to Economic Geography*, Oxford: Blackwell.

McKinnon, R. (1989), 'Financial liberalization and economic development: a reassessment of interest-rate policies in Asia and Latin America', *Oxford Review of Economic Policy* 5(4): 29–54.

Nelson, R. and Winter, S. (1982), *An Evolutionary Theory of Economic Change*, Cambridge: Harvard University Press.

North, D. (1990), *Institutions, Institutional Change, and Economic Performance*, Cambridge: Cambridge University Press.

O'Brien, R. (1992), *Global Financial Integration: the End of Geography*, London: Royal Institute of International Affairs.

Panitch, L. (2000), 'The new imperial state', *New Left Review* 2: 5–20.

Pempel, T. J. (1997), Regime Shift: Japanese Politics in a Changing World Economy, *Journal of Japanese Studies*, 23(2), 331–61.

Piore, M. and Sabel, C. (1984), The *Second Industrial Divide: Possibilities for Prosperity*, New York: Basic Books.

Pryke, M. and Allen, J. (2000), 'Monetized time-space: derivatives – money's 'new imaginary'?', *Economy and Society*, 29(2): 264–284.

Saxenian, A. (1994), *Regional Advantage: Culture and Competition in Silicon Valley and Route 128*, Cambridge: Harvard University Press.

Scott, R. (1995), *Institutions and Organizations*, Thousand Oaks: Sage.

SFC (2004), *A Special Note to Promote QFII*, Taipei: SFC (Securities and Futures Commission).

Skocpol, T., Evans, P. and Rueschemeyer, D. (1985), *Bringing the State Back In*, Cambridge: Cambridge University Press.

Storper, M. (1997), *The Regional World: Territorial Development in a Global Economy*, New York: the Guilford Press.

Storper, M. (2004) 'Society, community and economic development'. paper presented at the DRUID Summer Conference 2004.

Storper, M. and Walker, R. (1989), *The Capitalist Imperative: Territory, Technology, and Industrial Growth*, New York: Blackwell.

Thrift, N. and Leyshon, A. (1997), 'A phantom state? The de-traditionalisation of money, the international financial system and international financial centres', in A. Leyshon and N. Thrift (eds) *Money/Space: Geographies of Monetary Transformation*, London: Routledge.

Tickell, A. (2000), 'Dangerous derivatives: controlling and creating risks in international money', *Geoforum* 31: 87–99.

Wade, R. and Veneroso, F. (1998), 'The asian crisis: the high debt model versus the Wall Street-Treasury-IMF complex', *New Left Review* (March-April): 3–22.

Wade, R. (1990), *Governing The Market : Economic Theory and the Role of Government in East Asian Industrialization*, Princeton: Princeton University Press.

Wang, Zhen-huan (1996), *Who Ruled Taiwan? The Transforming State and Power Structure in Taiwan*, Taipei: JuLiu Press (in Chinese).

Weiss, L. (1998), *The Myth of the Powerless State* Ithaca: Cornell University Press.

Wu, Chyuan-yuan (1993), 'Financial liberalization reconsidered: a critique from economic sociology', *Taiwan: A Radical Quarterly in Social Studies* 15: 1–37 (in Chinese).

Yang, Zong-ren (2003), 'The underground futures business is still alive and well', *United News*, October 06, 2003 (in Chinese).

Yeh, Ming-feng (1997), 'A comparison between the of financial centers in Taiwan, Singapore and Hong Kong', *The China Commercial Bank Monthly* 16(10): 3–11 (in Chinese).

Zan, Ting-zhen (1993), 'The legalization process of Taiwan's futures trading service industry', *Taiwan's Securities* 38: 1–15 (in Chinese).

Zou, Zhen-yu (2002), 'The Characterization of Taiwanese State: A Perspective through the Construction of the Futures Trading Market', Masters Thesis, Department of Sociology, National Chengchi University (in Chinese).

Zysman, J. (1983), *Governments, Markets, and Growth: Financial Systems and the Politics of Industrial Change*, Oxford: Martin Robertson.

Chapter 11

Networked Governance for Global Economic Participation: The Case of New Zealand's Largest Service City

Steffen Wetzstein

Introduction

Since nineteenth century settlement, the livelihood and lifestyles of people in New Zealand have been intrinsically interwoven with relations to, and conditions of, overseas markets. A country similar in land size to Japan and the United Kingdom but with today's four million people 'by world standards a virtually empty country' (Clark and Williams, 1995, 21), New Zealand needs to engage in local-global interactions such as trading to overcome the economic disadvantages of being a small market. These processes have intensified over the last two decades under neo-liberal deregulation and the removal of protectionist borders. These political changes have created conditions for New Zealand people, firms and localities to become part of rapidly integrating global financial, trade and production networks (Le Heron and Pawson, 1996). Increasingly, territorial and industrial policy approaches are targeting on 'participation in globalization processes' as the way to secure profitability for local economic actors while simultaneously pursuing wider societal objectives.

Auckland is New Zealand's largest and fastest growing region. In 2001, it had a population of approximately 1.2 million, or 31 per cent of the national total. Current population growth averages 1.5 percent a year which is significantly higher than for New Zealand as a whole. The region is also an important part of the national economy; one third of New Zealand's businesses are now located in the Auckland region. The main contributor to Auckland's growth over the last decade has been the service sector, providing for the needs of the Auckland domestic market, other regions in New Zealand, and some export markets (*Market Economics*, 2002). Both its seaport and airport play a pivotal role in the regional and national economies, handling almost three quarters of the country's goods imports, and 40 per cent of exports (Joint Officials Group, 2003). Despite these figures, there are concerns about the increasing marginalization of Auckland nationally and globally (Le Heron and McDermott, 2001) as local development has shifted from production centred economic activity to a focus on local consumption, producer and consumer services. Moreover, with the internationalization of New Zealand's resource sectors even

Auckland's contribution as a business service hub for national firms is reduced. Thus, global participation has become a crucial policy challenge for the region.

In a competitive neoliberal institutional environment, collaborative responses to Auckland's global participation challenge are increasingly emerging. An example of a recent key initiative is the development of Auckland's Regional Economic Development Strategy (AREDS, 2002), which galvanised interests from central government, local government and local business to stimulate local-global economic interaction. Public and private policy practices increasingly include competitive and aspirational forms of benchmarking that help to construct imaginaries of globally spread city networks that incorporate Auckland. Under AREDS and in local government practice, Auckland has been benchmarked against Asia-Pacific cities with similar attributes in terms of size, trade function and language spoken. But how far can such an imagined Asia-Pacific region keep up with real connections between the local and the global? And more importantly, to what degree can policy and governance arrangements under contemporary neoliberal and globalizing conditions really influence a reconfiguration of spatial economic relations involving Auckland?

This chapter outlines shifts in the nature of governance arrangements and outlines an emerging silhouette of governance configurations on all geographical scales that have effects on the level of Auckland's inclusion in global economic circuits and networks. Governance is understood relationally, which means that coordination can only be influenced when links between actors are made.

The key claim is that governance arrangements affecting Auckland's global economic participation are increasingly resemble actor and resource networks, with context-dependent inclusion and exclusion of service activity interests. The remainder of this chapter is structured as follows: the next section provides an overview of contemporary understandings on urban and regional governance followed by an introduction to New Zealand's and Auckland's governance context. The next part discusses the rise of networked governing arrangements in Auckland. The chapter then critically evaluates the possibilities and limits of networking in governance contexts, and makes visible the emergent globalizing networked governance that affects the position of Auckland-based actors in economic processes. Concluding remarks point to new questions which may guide future thinking on the governance of globalising cities and regions.

Development Policies and Governance in Cities and Regions

Local Policy Responses for Globalizing Regions and Cities

Regions and cities have received increased attention as particular sites of economic coordination in a globalizing world. The 'new regionalist' writings (see Florida, 1995; Cooke, 1998, Storper, 1997) associate this trend with wider shifts in the nature of economic activity from labour-intensive production to knowledge and innovation rich work. These transformations are said to put a premium on territorially embedded social relations, actor networking and 'institutionally thick' local cultures as determinants of local growth. The extensive literatures on world- and global cities (for example Friedmann, 1986, Sassen, 1991, Taylor et al., 2002, Scott, 2001) describe and theorize the incorporation of particular cities into global structures of power and exchange based on their nature as command posts for the operations of multinational corporations and as centres of advanced services and information-processing activities. Crucially, such cities are also marked by deeply segmented social spaces and extremes in terms of poverty and wealth.

A range of policy responses to regional and urban development have emerged over the last decades which incorporate the 1960s Keynesian framework and 1980s neoliberal approaches. Conventional orthodoxy in economic and territorial development has been largely firm centred, incentive based, state driven, standardised and often centrally coordinated at the national level (Pike, 2004). A policy response emerging in the 1980s has been an urban-entrepreneurial approach based on speculative consumption-centred investment and the construction of place (Harvey, 1989). A key institutional arrangement within this policy framework is the 'public-private partnership' which promises to offer better returns on investment. Recently, Amin (1999) and others have emphasized the role of supply-side interventions that consider the facilitation of innovation, flexibility and actor collaboration as potential solutions to regional problems. This approach favours the mobilization of local resources and uses a very broad definition of what constitutes economic action incorporating non-economic factors such as 'Quality of Life'. Le Heron and McDermott (2001) however argue that under neo-liberal conditions, attempts to attract international capital through lowering the cost for doing business as well as strategies to encourage local businesses to develop attributes so that they can to link into the global economy have seldom proven successful.

Finally, local and global processes are connected through the concept of value chains which relate to earlier work on commodity chains (Gereffi and Korzeniewicz, 1994). For Le Heron and McDermott (2001) a 'global value chain' can be understood as the ability to assemble knowledge and capacity from disparate organizations across localities to produce globally demanded goods and services. Local specialization, rather than being based on the concentration of a particular skill or production base at a particular place, can be understood as contributions to international economic exchange. Local outcomes however are 'indeterminate, contextually specific, attainable through multiple pathways and nonlinear in nature' (368).

Regulating Urban and Regional Processes: Governance and Networked Institutions

Among many definitions, the term governance can be understood as the result of social-political-administrative interventions and interactions in a given policy field (Kooiman in Rhodes, 1996). Such goal-directed interventions or attempts to govern are performed in particular governance arrangements. From a political economic perspective, governance is best understood as a set of power relationships which determine whose interests are included or excluded in political and policy processes. Governance also denotes the increasing interdependent nature of coordinating economic activities across traditional domains of the state, business, non-profit organizations and other social actors (Jessop, 1997). Thus, it illustrates the intensification of societal complexity which flows from growing functional differentiation of institutional orders. In this context the state, understood as a series of social relations (Jessop, 1990), co-opts other interests to achieve governmental effects.

In the contemporary urban-regional development context, governance is centrally concerned with affecting the conditions for local accumulation in ways that allow economic growth and societal objectives to be reconciled. The framing and shaping of the conditions for local actors to participate in economic processes can be realised through the means of laws, regulations, incentives and disincentives as well through negotiation and interaction. The later relational forms of intervention are the basis for this analysis of the governance dimensions in Auckland's development. In this context, state – economy and state – business relations can be viewed at the heart of any governance problem, as development in capitalist societies is largely in the hands of private investors.

In New Zealand and United Kingdom contexts, business is often not included in public policy making arenas. This can, in part, be explained by a certain cultural difference between state and capital interests in both countries resulting in weak business representation on both political and policy levels (Perry, 2001, Valler et al., 2004). In contrast, the 'growth machine' concept states that US cities should be understood in terms of the efforts of property-owning elites to realize their interests in urban growth (Molotch, 1976). McGuirk (2000) examines the emergence of networked governing practices for urban renewal. She takes the intersection of central government property-led regeneration initiatives with local government planning regulation in Dublin as a forum in which to explore the multi-scaled policy networks (see also Bassett, 1996) constituting urban governance, and the role of local government within them. In McGuirk's account, local development interests and central government mechanisms became aligned in governing networks through which urban policy was produced and implemented. Elsewhere (2004) she theorizes urban governance from a wider perspective by connecting its practical accomplishment to questions of broader politico-economic embeddedness and to the territoriality of the state.

New Zealand and Auckland's Post-Restructuring Development and Policy Interventions

Economic Restructuring, Service Industries and Local–global Connections

New Zealand's export-focused economy has been built on the back of its primary base. Over time, ongoing economic diversification has been sought to secure new external sources of profits. For Easton (1997) 'the economic reforms of the 1980s are to be seen as a response to the new diversified political economy overriding the declining pastoral one' (49) and involving a radical shift from a long phase of protectionist policies and supportive intervention in New Zealand to a spatially open market-driven development model. Economic re-structuring of the 1980s and 1990s was largely driven by the central state as 'New Zealand has a highly centralized economy ... [and] only central government has mediated the performance of and opportunities for investors, local and from abroad, involved in local production' (Le Heron, 1987, 265). New conditions created by the 1980s neo-liberal reforms resulted in spatially divergent trajectories for economic sectors and regions in New Zealand. After-restructuring, recovery had been due to the rapid internationalization of the resource sectors as well as immigration-fuelled consumption-based urban development, particularly in Auckland. A third factor has been the rise of service activities in the economy. As Easton (1997) explains:

> Generally it is the service sectors which have filled in the gap left by the primary sector share decline. There are three factors here: First, primary and secondary industries use more services over time. Second, we have the tourist [services industry]. And third, services make up an increasing proportion of public and private consumption demand (139).

All three trends have been aided by a policy environment which has removed barriers to local-global interactions.

Auckland, the country's largest city and centre of the import substitution economy, had been particularly affected by the radical overhaul of the economy. For example, the region has been the biggest recipient of New Zealand immigrants over the last 15 years. Local development has been largely about the growth of the consumption side of the economy on the back of a rapidly growing service sector. As manufacturing employment declined by 1 per cent between 1991 and 2001, total employment increased by 34 per cent primarily through growth in business services, health and community services, wholesale trade, construction, retail and education (*Market Economics*, 2002). Often, these activities are performed in small businesses. Importantly, many of the jobs in services serve both Auckland and some of the demand from other regions as many service activities in other regions of New Zealand relatively declined due to changing business economics and the efficiency gains from electronic data capabilities. In addition, Auckland increasingly served as a control centre for firms operating in primary activities, in particular through the relocation of headquarter functions to the city-region. Auckland truly became New Zealand's service capital.

However, in recent years Auckland is said to have become increasingly marginalized from a local-global perspective. This trend is illustrated in Figure 11.1, which shows the high share of Auckland's economic activity that is sourced, or consumed, from within the region. These patterns reflect the high degree of self-sufficiency of the Auckland economy, and thus its global economic isolation. Le Heron and McDermott (2001) claim that the region is decreasing in importance even as a location for primary sector corporate functions as '... with the internationalization of New Zealand's resource sectors ... Auckland's contribution as a centre of international commerce, management and marketing is reduced' (370). Its traditional role as a physical gateway between New Zealand and the world also became fragile with other New Zealand locations re-positioning themselves favourably in regard to seaport mediated exporting activities.

Policy and Institutional Context for Auckland's Post-Restructuring Development

National policy and academic discourse focused on Auckland has appeared only in the last couple of decades. Previously, the city region's rapid growth and associated opportunities and problems had been largely ignored in the literature. Policy discourses in the 1960s and 1970s (Whitelaw, 1967 Bush and Scott, 1977) largely constructed the city-region in negative terms. Its particular status as New Zealand's largest city and locus of unparalleled growth threatened prevailing policy paradigms of spatially even growth patterns. During the neoliberal reforms, aspatial policy thinking had prevailed. Besides, in comparison to other regions Auckland and its actors were seen to perform quite well in adjusting to new conditions. Local development policy responses in a post-restructuring environment were largely of an entrepreneurial mode, centred on place promotion and property development (Murphy, 2003). However, increasing pressure on land and infrastructure led to the development of collaborative governance structures focused on growth containment policies at regional state level. Over recent years, those policy initiatives have expanded to incorporate economic development, transport policy and emerging urban design policy fields.

The neoliberal reforms changed the institutional environment for policy making and regional governance. Corporatized state actors introduced business practices and split policy making and delivery functions were coordinated on a contract basis. The delivery of many public goods was privatized. The central state moved away from representation in the regions (Moran, 1992). Increasingly, enabling local state decision-making was based on new forms of 'calculated' accountability. In 1996, MMP (Mixed Member Proportional form of parliamentary representation) – 'fundamentally a response to a new situation in which no two parties can fully encompass the politics of a much more diverse society and economy' (Easton, 1997, 49) – strengthened local representation in Parliament.

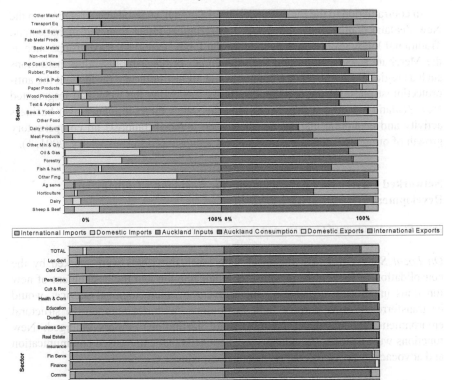

Figure 11.1 Significance of trade to the Auckland economy – manufacturing and service sectors

Auckland's government has always been marked by local fragmentation and patch protection as the city started off as a collection of market towns. Their sequenced consolidation created bigger local authorities that would often compete for scarce resources (Bush, 1977). A second local state layer was introduced in the form of a regional council responsible for implementing an environmentally framed, but development friendly, planning framework through the 1989 Local Government Act. Central government-local government relations had been difficult for a long time. These are partly due to the politically threatening size of Auckland's population and economy relative to the rest of the country, partly a result of a perceived cultural 'otherness' of the city-region and its people by non-Aucklanders.

In contradiction to the European corporatist model, state-business relations in the New Zealand context have emphasized government rather than self-management. Traditional lobby groups such as the farming lobby, export-focused groups such as the Merchant's Association and the often small business-representing institutions such as regional Chambers of Commerce had all lost influence throughout the anti-protectionist phase of the neoliberal reforms. Reflecting the weak and fragmented New Zealand business system 'even after the state's direct influence over business activity and economic management has been reduced, [almost] no compensatory growth of organized, encompassing interest groups has occurred' (Perry, 2001, 1).

Networked Governance Arrangements for Auckland's Globalizing Development

Intra State Networked Governance Arrangements

On Local State Level   On the local state level, the 1990s were marked by the consolidation of local government after amalgamation and the assignment of new functions under new legislation. The delivery of many regional services would be transferred to new purpose-built institutions. Within a competitive structural environment, an increasing complexity of institutional orders emerged. New functions were added to Auckland's local councils that emphasised communication and advocacy. A previous council manager explained:

> In 1989 local government in Auckland started as amalgamation, they focused on internal organization, administration. Since the mid-1990s they looked outwardly and identified issues for the city. There was a realization that they couldn't do things on their own ... they had to engage with other councils and central government as well as utility providers (Local government manager 1, 2004).[1]

However, problems facing the whole region proved difficult to tackle in a largely antagonistic local state culture. The shortcomings of fragmented decision-making and increasing pressures in areas of land-use, urban encroachment and heritage prompted local politicians to act in a more collaborative manner. Under more enabling conditions set out by central government, a range of forum's sprung up: the Mayoral Forum, the CEO-Forum, the Regional Growth Forum. While these arenas were described as almost dysfunctional in the beginning, increasingly local actors 'had to hold hands' in order to present a more unified interface to central government and to better coordinate local issues. Such developments, however, don't mean

---

1   During June 2003 and December 2004, a series of interviews was carried out with individuals from organizations involved in urban and regional governance in Auckland. Those interviewed included: key politicians, executives and policy professionals from local and central government, local business leaders from various organizational backgrounds, managers from economic development organisations and associations', regional funding and delivery agencies as well as private consultants and academics.

that the traditional hierarchical structures of the state somehow gave way to a new network-type of interaction. Rather, both forms of coordination became enmeshed in quasi-hierarchical structures whose exact configurations depend on the particular context. This complexity can be understood as follows:

Local and regional government in Auckland is a mix of both, a bit of hierarchy and a bit of network type institutional relationships. Networks have hierarchies within them, hierarchies of use. It's like the Google search engine – there is a dominance of certain websites which come up first ... the larger Councils are dominant because they have the resources to do the initiatives of regional significance ... so within the networks there is an unseen hierarchy of sheer scale (Local government manager, 2004).

Recently, the Local Government Act 2002 further enshrined both network and negotiation practices among the local state. Without ascribing particular roles and functions to local organizations, powers of competency were introduced as legal tools to re-organize the local division of state labour.

On Central State Level When a Labour-led coalition resumed power in New Zealand in 1999, it inherited a fragmented state machinery. Since then, institutional processes and practices have been re-worked under the programmatic umbrella of a 'joined-up-government'. Guiding this envisaged transformation was the 'State of the Centre Report' (State Services Commission, 2001) which, commissioned by the Prime Minister Helen Clark herself, reviewed existing governing practice. It pointed to the need to rebuild the government interface with its citizens around a healthy balance between outcomes, outputs and capability. In order to improve alignment, the advisors suggested the establishment of networks of related agencies at both the centre and in the regions to better integrate policy, delivery, and capability building. While such networks might be comprehended as less rigid and less hierarchical in their internal power distribution than previous top-down arrangements, the locus of control actually remains largely unchanged as '... network membership is compulsory [and] component organizations could not opt out' (26). Since the publication of this report, governing practices appear to have been altered to some degree:

Previously you consult with them [other departments and agencies], which meant drafting a discussion document, sent it out and getting a submission back. Today, there is much more face-to-face interaction, we have joint officer groups and make more use of advisory groups. It's a shift from consulting to engagement (Central government manager, 2004).

Part of the reorganization of the central New Zealand state is a move back into the regions. Even with modern communication technology and faster transportation, 'connecting at a distance' is not sufficient enough for effective governing. Increasingly, actors co-locate. For example, the current CEO of a government economic development agency – who in the recent past would only lead from his Wellington headquarters – is now represented simultaneously in offices in Wellington and Auckland (New Zealand Trade and Enterprise, 2005). And the University of Auckland considers it essential to have an institutional representation in Wellington, because 'government happens' in New Zealand's capital (Academic, 2005). On the one hand these co-location trends highlight the importance of proximity in spatial interactions (Sorenson, 2003, Porter, 2003, Brown and Barnett, 2004), on the other hand it shows the importance of state personnel as a key actor in a 'relational state' that actively seeks to shape local governance trajectories (Jones et al., 2004).

Between Central and Local State Actors   Networking between central and local state institutions has been enhanced under the wider political project of 'partnerships'. While this discourse has been circulating in Anglo-Saxon policy circles for a number of years (DETR, 1997, in Valler et al., 2004), in New Zealand it was under the newly elected government in 1999 that sustained efforts to formalize partnerships between national government, local state institutions and communities begun to guide political and policy practices. These can be linked to 'integrated policy making', a theme which had become increasingly salient within central and local government in the UK (Betteley and Valler, 2000). In New Zealand, the 'partnership' aspiration was not a sole ambition of central government alone; this impulse was shared by the local and regional layers of government throughout the country. A policy professional explained that the:

> ... partnership model is the alignment of two ambitions. Central government asked how we can do things differently in order to better deliver locally, and local government in general realized that by far the biggest piece of local expenditure is under the control of central government' (Local government manager 2, 2004).

A very recent example of the cross-fertilizing nature of 'joined-up-government' is the way one of the fractions of local government in Auckland responded to new conditions of engagement. In a recent report, the Auckland Regional Council's strategic policy team envisaged that Auckland's local authorities may be able to better co-ordinate their dealings with central government on major policy and priority issues through a regional grouping which would focus on the alignment of broad policy outcomes and associated expenditure (Auckland Regional Council, 2004). It is interesting that the idea of establishing such a structure to deal with central government is becoming a more naturalized response to new governing conditions. It seems to be driven less by perceptions of crisis, but rather by emerging institutional opportunities for more effective governance models.

State-Business Networked Governance Arrangements

New Zealand's Research and Innovation Policy Field   Over the last couple of years, proliferating state-business relations have often taken the form of networked institutional arrangements. They involve both the central and the local state level as well as particular business interests, dependent on the policy context. In the research and innovation policy field, a key example is the so-called 'Knowledge Wave' project initiated from the University of Auckland, the country's largest tertiary education institution, which was looking for ways of increasing funding opportunities. It grew into a nationwide advocacy campaign for the promotion of the knowledge-based society, highlighted by two high-profile conferences in Auckland in 2001 and 2003. The project was supported by mostly-Auckland based educators and business leaders. These were initially activated *via* old corporate networks through the University's Vice Chancellor John Hood, and later through the formation of new networks of younger, more public good spirited corporate leaders. Big businesses whose future depends on the overall economy such as telecommunications, financial and global management consultancy interests participated as sponsors.

This initiative subsequently contributed to new national policies in the form of the Growth and Innovation Framework (GIF) (The Office of the Prime Minister, 2002). As to the attributes of this kind of state-business interaction it was commented that:

> [b]usiness was asking for intervention, government officials went and listened. Government was able to establish a dialogue. The most important thing which came out of it was the developing of a shared understanding of what the roles of government and business were' (Central government policy analyst, 2004).

Closer state-business relations in policy development can also be illustrated by the commissioning by the Prime Minister/Cabinet Ministers of two very influential reports on economic issues facing New Zealand (Boston Consulting Group, 2001, LEK Consulting, 2001). The co-opting of globally connected private sector knowledge networks by central government contributed markedly to the development of the GIF policies. From a theoretical perspective, these examples show the co-dependency of state and business in contemporary governance settings.

Auckland's Regional Economic and Urban Development Policy Fields In the past, local business had only been represented by the traditional membership-based service institutions such as the regional Chamber of Commerce. Around the year 2000, increasing relocations of corporate businesses to Australian cities – reflected in increasing vacancy rates in the inner city commercial property sector – mobilized a local investor to political action (Business leader 1, 2004). Other business leaders joined a campaign under the rubric 'Competitive Auckland', representing a mix of mostly development, consultancy and educational interests. It aimed both to raise awareness of Auckland's globally referenced economic underperformance and to influence state institutions to promote Auckland globally.

This autonomous mobilisation of business interests became part of the founding platform for Auckland's Regional Economic Development Strategy (AREDS), a public and private sector crossing policy and governance experiment that produced a widely applauded 20 year strategy for Auckland's global economic connectivity. Other founding influences were a new collaborative moment in local government's economic development community – borne out of loosing a high-profile overseas investment opportunity – and the re-emergence of the central state in regional economic development planning. AREDS went through three distinct phases: a strategy development process, interim implementation arrangements and today's governance structures under the Auckland Regional Council. The first phase was guided by networked leadership consisting of key representatives from various regional interests – a process funded by central government grants and some in-kind contribution from local government. In the second phase, informal governance arrangements centring on the Auckland Regional Council co-existed with a very narrowly defined, formally governed, central state-framed strategy implementation process that was often criticized for its inadequacy for dealing with the management of globalising economic development work. Today's institutional framework consists of more networked governance at the strategic level (with some private sector involvement), complemented by a service-focused regional trust in charge of investment attraction, regional promotion and facilitation of regional projects.

AREDS can be understood as a networked governance arrangement involving central state, local state and business interdependencies – each representing its own networked interests. These actor networks brought a particular mix of resources to this initiative: the central state used legislative and policy-based funding powers, the local state contributed office space, project management support and operational resources, and business added value through particular business relevant knowledges as well as access to business leader networks. This type of governance is different to past regional arrangements as 'AREDS is much more a network thing [in contrast to] the Regional Growth Forum [which] is much more of a hierarchical thing' (Local government manager 1, 2004).

In 2003, Competitive Auckland became 'Committee for Auckland', a:

> private sector, non-profit organization where members' skills, resources, enthusiasm and considerable influence ... will contribute to the economic, social and environmental wellbeing of New Zealand's largest urban area' (Committee for Auckland, 2003).

This business initiative moved from an advocacy group to an initiative which works with corporate actors in Auckland facilitating projects on economic and urban revitalisation. This transformation signals a shift in policy focus to issues of urban development, liveability and sustainability in New Zealand. This is revealed by the emergence of new organisations such as the Government Economic and Urban Development Office and Urban Design panels within the Auckland region. The transformed business initiative can be seen as part of an emerging third-party institutional infrastructure in Auckland which, by acting as intermediaries between public and private sectors organizations, simultaneously pursues economic and wider regional objectives. In common with other contemporary institutions operating across this interface, it engages in temporary projects rather than on a permanent basis. Table 11.1 shows some key networked governance arrangements which have emerged in, or have affected, Auckland's regional development over the last decade.

Evaluating Networked Governance Arrangements: Making and Missing Connections

From a relational perspective, a possible way to evaluate the success or failure of networked governance arrangements is to measure the number and the depth of connections made between key actors. Under AREDS, particular business

Table 11.1   Networked governance arrangements for Auckland's development

Year	Local state	Business	Central state
Early 1990s	Mayoral Forum, CEO-Forum		
1999	Regional Growth Forum		
2001	Auckland Regional Economic Development Group	Competitive Auckland; Knowledge Wave	Joint-Up Government Government
2002	Auckland Regional Economic Development Strategy		Growth and Innovation Framework; Tertiary Education Commission
2003	Auckland Transport Action Group Committee for Auckland		
2005	Auckland Regional Economic Development Forum		

Source: Author.

interests contributed to making this initiative work even though wider private sector representation and connections have been difficult to achieve. While 'Competitive Auckland' took part in this initiative, engagement with more traditional business institutions was sub-optimal (Business leader 2, 2004). Connections to the corporate world were largely absent. Auckland's universities were involved, but during implementation it became apparent how competitive the institutional structures in the tertiary education industry were, and how difficult it was to achieve agreement between them.

Given the ambition of AREDS to influence the regional economy, networked governing attempts had limited reach into private sector investment processes; only a limited number of connections were made. For example, big infrastructure organisations such as the sea- and airports were not directly affected by the work of AREDS (Business leaders 2 and 3, 2004). Equally, no groups of small investors were mobilised in any significant way. While some service business interests such as management consultancies played a role in the AREDS knowledge production processes, the great majority of Auckland's business and consumer service firms were not involved in AREDS. Close observers readily admit that despite new institutional arrangements 'there is no investment change on the ground' (Business leader 1, 2004). However, the thirty biggest firms in Auckland have recently been approached in an effort to better understand the location decisions of 'big business' for public infrastructure provision planning.

The creation of deeper connections which would allow opportunities for altered investment behaviours were challenged by institutional barriers between the public and private sector. These are largely constituted by the traditionally weak New Zealand business system, an individualistic local culture and incompatibilities between public and private sector work modes. A policy professional describes the latter:

> The fundamentals haven't changed. Government has a long-term focus, 50 years, 30 years horizon. Business thinks short-term, maximum five years. Our languages are totally different. What drives the public sector is political consideration. The private sector is maximising resource efficiency. The public sector challenge is to translate its own views into the language of business, into short-term economic language, so that the private sector can understand (Local government policy analyst, 2004).

However, 20 years after the onset of the neoliberal reforms in New Zealand, more people on the interface between public and private sectors are now comfortable with continuum-thinking rather than box-thinking (Local government manager 1, 2004).

Finally, while it would be easy to dismiss AREDS on the grounds of lacking public-private engagement, if one assesses capacity building for Maori – the indigenous people – Pacific Peoples and the migrant community, evaluation can be more sympathetic. Despite considerable engagement challenges, '... under AREDS, Mana Whenua [the local Maori people living in the wider Auckland area] came together for the first time in the region' (Local government manager 3, 2004) and new migrants were given a voice in governance for the first time after the 1990s immigration boom.

Emergent Globalizing Networked Governance Affecting Auckland

Multi-scalar Networked Governance

Auckland's recent history of economic governance highlights the importance of networked governance arrangements, rather than purely hierarchical or market forms of coordination. In the earlier phase of neo-liberal transformation, local state networks had been the emergent, unplanned answer to the gap left by central government and the shortcomings of market arrangements. Since the mid-nineties, more experimental and deliberately networked forms of governance have been developed as tools for governing regional and economic processes. These processes have broadened to include both central state institutions and particular interests among the local business sector. The individual site and actor-specific resources which these networks bring to governance projects create in their interaction synergies and interdependencies that increase the likelihood of desired governance effects. Importantly, network emergence at different institutional sites has been non-linear in time as it is contingent on the particular structural, institutional and political conditions.

Networked governance arrangements are not confined to actors and processes operating on local, regional and national scales. Rather, they can be understood as part of multi-scalar networked governance that includes emerging arrangements on all geographical scales which affect Auckland's regional and economic processes. A wide range of actors is included: state networks, local and global business networks such as 'Competitive Auckland' and the New Zealand expatriates association 'KEA', as well as more informal networks such as transnational migrants. All contribute resources for Auckland's and New Zealand's global economic participation.

Table 11.2   Multiplicity of networked governance arrangements on all geographical scales with effects for Auckland's development

Geographical scale/ institutional site	Central state	Local state	Business
Local	Local economic development grants	Economic development agencies; Public-private partnership for indoor Arena	Public-private initiatives (Committee for Auckland)
Regional	Regional partnership programme; Government Economic and Urban Development Office	Auckland Regional Economic Develop-Forum/Business Unit	Regional economic development planning initiative (Competitive Auckland)
National	Growth and Innovation Framework (GIF)	Advocacy for Auckland's political representation at central government (co-location)	Private consultants' input into central government policy
Global	NZTE-network	Sister city links	KEA-Expat-network

Source: Author

Crucially, governance is increasingly about the activation of such multiple and multi-scalar resource networks, involving institutional experimenting and reflexivity. From a regulatory perspective, AREDS shows that governance effects often take on discursive rather than material forms as actors' assumptions about, and consequent attitudes towards, regional and economic processes change rather than actual investment processes. Changed actor assumptions are now based on less precise and more integrated understandings of what is supposed to be governed. This domain is now expanding from traditional economic objects such as firms and markets to include social and cultural aspects. Thus, new 'outcome-based' conceptions of economic development incorporate transport, education and 'quality of life' issues. Table 11.2 shows examples for networked governance arrangements with effects for Auckland in regards to the geographical scale of their emergence and the institutional sites they are located in.

Up-scaling of Governance and Policy Objectives

Observations from Auckland also show another dimension of globalizing networked governance: the ongoing attempts to up-scale governance arrangements. Table 11.3 illustrates these tendencies for particular intervention fields. It shows that during the last decade some arrangements have moved up in scale while others have emerged anew or re-emerged on the national scale. For example, firm support has been expanded from a function of the local state in the form of local economic development agencies to include particular support arrangements provided by a central state agency, New Zealand Trade and Enterprise. The latter exclusively targets firms producing for export markets in the Biotechnology, Information and Communication Technologies, and Creative Industry sectors. Up-scaling is also reflected in policy objectives, for example in the case of regional economic development policy. Because 'the scale envisaged by regional policy in New Zealand is much smaller than that generally referred to in the international literature' (Schoellmann and Nischalke, 2005, 102), a recent policy review of the central government regional partnership programme suggested an expansion of the policy focus of 'partnering' from local organisations to regions. Crucially, institutional impulses to up-scale the spatial horizons of governance come not just from state, but also from business interests such as 'Competitive Auckland'. Both, the constant up-scaling of networked governance arrangements and their emergence in multi-scalar settings can be viewed as indicators of globalising networked governance that affects Auckland's development.

Key Role of the Central State

Despite conceptualizing networked governance as a key form for attempts to manage urban and regional economies under conditions of globalization, the Auckland example shows that only the central state possesses the key means to intervene in regional and economic processes. Its techniques feature a mix of facilitative and supportive mechanisms that are now customised to different economic sectors and

Table 11.3   Up-scaling of institutional governance arrangements in key policy fields for Auckland's development over the last decade

Time Period	1990s			2000s		
Policy field/ intervention	Policy focus	Institutional arrangement	Scale of intervention	Policy focus	Institutional arrangement	Scale of Intervention
Place Marketing	Local brand image	Local economic development agencies; Local councils	Local	Single facilitation point	Regional Economic Development Unit	Regional
Firm Support	Business support	Local economic development agencies	Local	Small business support	Local economic development agencies	Local
				Growth and Innovation Framework (GIF)	New Zealand Trade and Enterprise	National
Education	City employment	Local councils	Local	Regional labour market analysis	AREDS	Regional
				National Tertiary Education Strategy	Tertiary Education Commission	National

Source: Author.

industries depending on perceived growth potentials. But neo-liberal restructuring has resulted in a smaller and institutionally fragmented New Zealand state with limited resources to directly affect economic changes. From a relational viewpoint, this has two implications for governance. First, current re-connecting of state actors can be seen as a precondition for maximizing the potential to intervene economically. Second, the co-opting of business and other societal interests and their particular multi-scalar resource networks is not just vital but absolutely necessary to achieve influence. These trends show that state and business are co-dependent. Neither the fragments of the state nor business actors have, in isolation, the answers as to how to manage territorial processes under globalizing influences. In this context, networked forms of coordination potentially open up more space for dialogue, negotiation, and individual commitment between actors. This offers hope that urban and regional policy problems may be tackled more effectively than in the past.

Auckland's Contested Role in Connecting Local and Global Processes

Recently emerged policy discourses and practices in New Zealand highlight the importance placed on Auckland as a key site connecting local and global processes. Thus, governance arrangements in a range of national policy fields show trends towards the increasing inclusion of actors from New Zealand's largest city. In this context, Auckland can be considered the most fruitful space for governing attempts to influence local actors to become part of global economic processes. Auckland features prominently in this governance moment because of its size, concentration of knowledge, technology and labour. There are numerous examples of the region's local – global connections including the airport and seaport links, transnational migrant networks, high-technology export firms, and global knowledge and entertainment circuits.

But Auckland's current repositioning in policy discourse and practice has limits. In New Zealand, central government will always tend to pursue policies aimed at spatial equality, making it highly unlikely that one region will receive favourable treatment for long, at least openly. Thus, Auckland will, for political reasons, probably always fall short of its potential in regards to government support, even under conditions of globalization. Viewed from this perspective, the city-region may find it harder to compete internationally. Ironically, one might say the city-region is 'too big for New Zealand, but too small for the world'. Furthermore, in a neoliberal environment, governance decisions and arrangements are subject to, and associated with, often contradictory and competing political and politicized projects. For example, while ownership of Auckland's seaport was transferred to the Auckland Regional Council in 2004, just two years previously Auckland's key local council had sold its shares of the airport. Strategic and integrated management of key assets of Auckland's globalizing economy had been hindered by the competing political aspirations that ruled on the day. Therefore, Auckland's perceived potential to connect New Zealand to the world may be overrated, its increasing policy attention contested, and its favourable governance status probably short-lived.

Conclusion

Using the example of Auckland in New Zealand, this chapter argues that under new conditions of neo-liberalized institutional relations and globalizing economic integration, experimental and networked forms of urban and regional governance are emerging. Those arrangements are based on asymmetrical actor co-dependency with a strong influence exerted by the central New Zealand state. Despite proliferating mobilization of business interests, the central state remains the main policy actor able to shape – at least to some degree – the conditions for urban and regional development. Central state interests are part of an emerging globally stretched networked governance architecture, which have effects on regional and urban processes in Auckland. This analysis shows that governing capitalist development remains a challenge, and constantly poses new policy problems. There is much to commend in Le Heron's interpretation that 'the two-way traffic interactions between New Zealand territory and other places are increasingly about attempts to activate resources, ensuring they are valued in global circuits and networks' (1996, 15). In fact, this is what governance of development processes is mainly about in the New Zealand and Auckland case: the activation of multiple and spatially diverse resource networks. From a relational viewpoint, networking to connect dispersed resources to achieve such interactions and connections is the primary way in which governance is achieved.

But governance comprehended in this relational way also highlights new problems. Networks of governing actors will always include and exclude particular groups in society, and therefore lead to inequalities. Not everyone can participate in global processes. For example, 'non-exporters' as well as many service oriented firms such as those in the tourism and property sectors currently miss out on favourable incentives from central government. Auckland's sizeable small business sector is largely ignored too. There is also a gap opening between processes of exclusionary global economic participation and closer local participation in decision-making that leads to new political and representation struggles. Difficulties of public-private engagement remain serious, as decision-making frameworks, objectives, languages and resources differ markedly between these sectors. And from a spatial perspective, Auckland's primacy as a key connecting site between the local and the global is not given, but always contested.

Finally a remark on what networked governance for global economic participation of New Zealand's largest city – and key service centre – might mean in the future. Auckland is part of a New Zealand economy that is small compared to the world, located at the global periphery and increasingly diversifying. Auckland's economy is domestically and consumption oriented. Thus, intentional connections between local and global economic processes will not be easy to achieve. Facilitation will likely be customized. Tapping into, building and maintaining particular networks will therefore continue to be an important aspect of economic governance. In fact, governance for a globalizing Auckland thought along 'relational lines' would mean creating and stitching together resource networks which can influence the insertion of local actors into global circuits and chains. Crucially, this would include a focus on policy drivers that support global value chain connections and development. In this context one can ask: can 'globally connecting' local infrastructure such as the privately owned and operated airport in Auckland be influenced? The answer is uncertain. The challenges to governing capitalist development processes under globalizing conditions are enormous. Thus, the nature and extent of Auckland's incorporation in the Asia-Pacific region and the wider global economy is by no means obvious!

Acknowledgements

I greatly acknowledge the financial support of the University of Auckland as well as the Travel Award of the IGU Commission on the Dynamics of Economic Space that allowed me to attend the Birmingham Meeting of the IGU in 2004. I am very grateful to my dissertation supervisors Professor Richard Le Heron and Associate Professor Larry Murphy of Auckland University for their overall guidance and professional support.

References

Academic (2004), interview with author, 17/03/2005.

Amin, A. (1999), 'An Institutional Perspective on Regional Economic Development', *International Journal of Urban and Regional Research*, **23**, 365–378.

AREDS (2002), 'Growing Auckland: Auckland Regional Economic Development Strategy 2002-2022', www.areds.co.nz/.

Auckland Regional Council (2004), 'Progressing the local/central government relationship', report to Auckland region strategic directors meeting, 11/11/2004, unpublished document.

Bassett, K. (1996), 'Partnerships, Business Elites and Urban Politics: New Forms of Governance in an English City?', *Urban Studies*, **33**, 539–555.

Betteley, D. and Valler, D. (2000), 'Integrating the economic and the social: policy and institutional change in local economic strategy', *Local Economy*, **14**, 295–312.

Boston Consulting Group (2001), 'Building the Future: Using Foreign Direct Investment to Help Fuel New Zealand's Economic Prosperity', Auckland: www.executive.govt.nz/minister/clark/innovate/bcg.pdf; ... bcg-2.pdf.

Brown, L. and Barnett, J.R. (2004), 'Is the corporate transformation of hospitals creating a new hybrid health care space? A case study of the impact of co-location of public and private hospitals in Australia', *Social Science and Medicine*, **58**, 427–444.

Bush, G. (1977), 'Auckland: its Government and Misgovernment', in G. Bush and C. Scott (eds), *Auckland at Full Stretch: Issues of the Seventies* Auckland: Auckland

City Council and Board of Urban Studies, University of Auckland, pp. 238–253.

Bush, G. and Scott, C. (1977), 'Auckland at full stretch: issues of the 1970s', Auckland: Auckland City Council and Board of Urban Studies, University of Auckland.

Business leader 1 (2004), interview with author, 10/11/2004.

Business leader 2 (2004), interview with author, 19/11/2004.

Business leader 3 (2004), interview with author, 06/12/2004.

Business leader 4 (2004), interview with author, 08/12/2004.

Central government manager (2004), interview with author, 22/11/2004.

Central government policy analyst (2004), interview with author, 22/11/2004.

Clark, M.S. and Williams, A. (1995), *New Zealand's Future in the Global Environment? A Case Study of a Nation in Transition*, Wellington, N.Z.: GP Publications: New Zealand Employers' Federation.

Committee for Auckland (2003), www.competitiveauckland.co.nz, 27/08/2003.

Cooke, P. (1998), 'Introductions. Origins of the Concept', in H.-J. Braczyk (ed.), *Regional Innovation Systems*, London: UCL Press, pp. 2–25.

Easton, B.H. (1997), *In Stormy Seas: The Post-war New Zealand Economy*, Dunedin, N.Z.: University of Otago Press.

Florida, R. (1995), 'Toward the Learning Region', *Futures*, 27, 527–536.

Friedmann, J. (1986), 'The World City Hypothesis', *Development and Change*, 17, 69–83.

Gereffi, G. and Korzeniewicz, M. (1994), *Commodity Chains and Global Consumption*, Westport, CT: Greenwood Press.

Harvey, D. (1989), 'From Managerialism to Entrepreneurialism: The Transformation in Urban Governance in Late Capitalism', *Geografiska Annaler*, 71 B, 3–17.

Jessop, B. (1990), *State Theory: Putting Capitalist States in their Place*, Oxford: Blackwell.

Jessop, B. (1997), 'The governance of complexity and the complexity of governance: preliminary remarks on some problems and limits of economic guidance', in A. Amin and J. Hausner (eds), *Beyond Market and Hierarchy*, Cheltenham: Edward Elgar Publishing.

Joint Officials Group (2003), 'Auckland Transport Strategy and Funding Project: Final Report', http://www.treasury.govt.nz/release/jog.

Jones, R., Goodwin, M., Jones, M., and Simpson, G. (2004), 'Devolution, state personnel, and the production of new territories of governance in the United Kingdom', *Environment and Planning A*, 36, 89–109.

Le Heron, R. (1987), 'Rethinking Regional Development', in P.G. Holland and W.B. Johnston (eds), *Southern Approaches: Geography in New Zealand*, Christchurch/NZ: New Zealand Geographical Society.

Le Heron, R. and Pawson, E. (1996). *Changing Places. New Zealand in the 1990s*, Auckland: Longman Paul.

Le Heron, R. (1996), 'Introduction', in R. Le Heron and E. Pawson (eds), *Changing Places. New Zealand in the 1990s*, Auckland: Longman Paul, pp. 1–19.

Le Heron, R. and McDermott, P. (2001), 'Rethinking Auckland: Local Response to Global Challenges', in D. Felsenstein and M. Taylor (eds), *Promoting Local Growth*, Aldershot: Ashgate, pp. 365–386.

LEK Consulting (2001), 'New Zealand talent initiative: strategies for building a talented nation', www.executive.govt.nz/minister/clark/innovate/lek.pdf.

Local government manager 1 (2004), interview with author, 09/12/2004.

Local government manager 2 (2004), interview with author, 19/11/2004.

Local government manager 3 (2004), interview with author, 10/11/2004.

Local government policy analyst (2004), interview with author, 10/11/2004.

Market Economics Ltd (2002). Linkages within and between economies. Auckland: Report for AREDS, www.areds.co.nz/ Doc_Library/Linkages_within_&_between_Economies.doc.

McGuirk, P. (2000), 'Power and policy networks in urban governance: local government and property-led regeneration in Dublin', *Urban Studies*, **37**, 651–672.

McGuirk, P.M. (2004), 'State, strategy, and scale in the competitive city: a neo-Gramscian analysis of the governance of 'global Sydney'', *Environment and Planning A*, **36**, 1019–1043.

Molotch, H. (1976), 'The city as a growth machine: toward a political economy of place', *American Journal of Sociology*, **82**, 309–332.

Moran, W. (1992), 'Central and Local Government Reform and Interaction', in S. Britton, R. Le Heron and E. Pawson (eds), *Changing Places in New Zealand. A Geography of Restructuring*, Christchurch: New Zealand Geographical Society, pp. 225–230.

Murphy, L. (2003), 'Remaking the city: property processes, planning and the local entrepreneurial state', in A. MacLaran (ed.), *Making Space: Property Development and Urban Planning*, London: Arnold, pp. 173–193.

New Zealand Trade and Enterprise (2005), http://www.nzte.govt.nz/section/11904.aspx, 10/06/05.

Perry, M. (2001), *Shared Trust in New Zealand – Strategies for a Small Industrial Country*, Wellington: Institute of Policy Studies.

Porter, M.E. (2003), 'The Economic Performance of Regions', *Regional Science*, **36**, 549–578.

Pike, A. (2004), 'Heterodoxy and the governance of economic development', *Environment and Planning A*, **36**, 2141–2161.

Rhodes, R. (1996), 'The new governance: governing without governance', *Political Studies*, **44**, 652–667.

Sassen, S. (1991), *'The Global City: New York, London, Tokyo'*, New Jersey: Princeton University Press.

Scott, A.J. (2001), 'Globalization and the Rise of City-regions', *European Planning Studies*, **9**, 813–826.

Schoellmann, A. and Nischalke, T. (2005), 'Central Government and Regional Development Policy: origins, lessons and future challenges', in J. Rowe (ed.), *Economic Development: The New Zealand Experience*, Ashgate: Aldershot, UK, pp. 75–108.

Sorenson, O. (2003), 'Social networks and industrial geography', *Journal of Evolutionary Economics*, **13**, 513–527.

State Services Commission (2001), report of the advisory group on the 'Review of the Centre', http://www.ssc.govt.nz/display/document.asp?NavID=177&DocID=2776.

Storper, M. (1997), 'Regional Economies as Relational Assets', in R. Lee and J. Wills (eds), *Geographies of Economies*, London: Arnold.

Taylor, P.J., Walker, D.R.F., Catalano, G. and Hoyler, M. (2002), 'Diversity and power in the world city network', *Cities*, **19**, 231–241.

The Office of the Prime Minister (2002), 'Growing an innovative New Zealand', Wellington: www.beehive.govt.nz/innovate/innovative.pdf.

Valler, D., Wood, A., Atkinson, I., Betteley, D., Phelps, N., Raco, M. and Shirlow, P. (2004), 'Business representation and the UK regions: mapping institutional change', *Progress in Planning*, **61**, 75–135.

Whitelaw, J.S. (1967), *Auckland in Ferment*, Auckland: New Zealand Geographical Society.

Chapter 12

Lowering the Professional Frontier: A Service Market in Development in Ahmedabad, India

Harald Bekkers

Introduction: Brokers at the Local-global Interface

A market is governed by a *single* standard of competitiveness that is determined by the most competitive product in that market. The competitive edge of this product is based upon a particular combination of price, quality and service, and can be met or beaten through a different combination of characteristics, a different strategy: to be a discounter, a service champion or just a better balancer of all the basics. For this reason there is nothing more threatening for an entrepreneur than when a new player enters the market with a different business model as this 'upsets' the established relationship between price and quality and services and thereby sets a new standard to which all market participants must somehow adjust (Fligstein, 2001). This is exactly what happens with economic globalization: an exogenous, global standard of competitiveness starts to challenge existing, local, indigenous practices. Then the issue of *compatibility* arises: how can this new standard be met? This pressure for compatibility translates again into a need for *convergence*, to 'copy globally in order to cope locally'.

Pressures for institutional compatibility and convergence work on four categories of economic institutions:

1. At the macro-level one can distinguish institutions that, following North (1985; 1990), reduce the transaction costs of exchange for a given product within the market place and thereby contribute to the creation of wealth. Such institutions are property rights, governance structures and exchange regimes.
2. At the meso-level one can distinguish institutions that, following Porter's (1990) theory of 'competitive advantage', contribute to the creation of new production factors such as specific knowledge, management tools and technologies. Such institutions are 'clusters' or, more generally, inter-firm business networks, consisting of firms, associations, research institutions, banks and (local) government.

3. Descending further from the market to the firm, another category of institutions that can be distinguished are concepts and tools to improve business management such as 'just–in–time' production and 'total quality management' (Bessant, 1991, Best, 1990).

4. Finally, along with institutions that *increase competitiveness* and *competition*, Fligstein distinguishes 'conceptions of control': institutions that *decrease (local) competition* and create a degree of order and stability in a particular market (2001:33). Such institutions are of particular importance in times of instability, when new markets are being formed or when existing markets are being invaded by external competitors, as is the case with economic globalization (2001:35).

Institutional compatibility and convergence should not be confused with a straightforward process of global homogenization. Rather, this is an argument for *context-based modernization*, for *local solutions based upon global solutions*. Institutional compatibility and convergence refer to a balancing act between exogenous market institutions and best practices and indigenous business culture without lapsing into a celebration of indigenous entrepreneurship or relapsing into modernization theory. *How* these institutions and best practices are blended with local culture and creativity should be understood as a process that actively needs to be *brokered* at the *local-global interface*, where two systems, one local, the other more global, each with its different standards, practices and institutional set-up, meet (Bekkers, 2005, Bekkers, in preparation). In Ahmedabad, a secondary-level city of 4.5 million inhabitants in Western India, predominantly small local 'management consultants' play this important brokerage role.

Ahmedabad is the economic capital of Gujarat, one of India's most industrialized states. Although it lacks the metropolitan flavour of Mumbai or New Delhi, Ahmedabad is a modern city by Indian standards. Toward the end of the nineteenth century it gave birth to a Fordist textile industry and, hence, came to be known as 'one of the few examples of indigenous capitalist industrialization in the colonial Third World' (Lakha, 1988:1, see also Oonk, 1998, Spodek, 1965, 1969). In the 1980s the city suffered badly from widespread bankruptcies in the textiles sector, but also showed resilience with an explosion in the number of small-scale industries (SSI) accompanied by economic diversification into chemicals, pharmaceuticals and consumer services.[1] Overall, however, Ahmedabad appears to be a city 'out of control', squeaking on the pressures brought about by India's liberalization process. The city is characterized by intense economic competition (up to the point that it fosters economic criminalization), social competition (up to the point that it turns into bloody caste conflicts), a decline of public developmental institutions (up to the point that they have become all but irrelevant for Ahmedabad's economic development); and a kind of 'de-linking' of the professional institutions that have

---

1   The number of SSIs has risen from 10,919 in 1980 to 29,661 in 1990 to 58,332 in 2000 (Government of Gujarat (2001:S59).

become world-class (up to the point that they have also become all but irrelevant) (Bekkers, 2005, Bekkers, in preparation, Spodek, 1989, 2001).

In this city out of control there is little to be found in terms of formally organized business networks or service markets. Instead, Ahmedabad harbours what could be labelled an 'informal service configuration', which is held together by local consultants, for two reasons:

1. Consultants broker *vertically*. Consultants, more than other professionals, are willing and able to *lower the professional frontier*, develop new markets, and introduce professional best practices to not yet professionalized family firms. In the process they can be found tinkering, for instance, with Japanese sun greeting ceremonies to make them fit in to Ahmedabadi companies so that better collegial collaboration will lead to higher quality production (thereby struggling with such questions as to what extent low-caste cleaners should be included in the new community set-up).
2. Consultants broker *horizontally*. Consultants network with fellow consultants and institutions. In doing so, they weave together Ahmedabad's otherwise fragmented business services infrastructure and provide informal coordination and synchronization where more formal governance is lacking.

Nevertheless, despite playing these important brokerage roles, with regard to Ahmedabad's economic development the role of local consultants remains rather ambiguous. This chapter will explore and explain this paradox. Section Two will describe how consultants work to *lower the professional frontier* and broker between the professional and entrepreneurial domain. Section Three will describe how the *reputation game governing the market* for consultancy services nevertheless produces a gap between consultants and entrepreneurs. Section Four will, given these two different stories, draw conclusions related to this service market in development.

Management Gurus, Local MBAs and Familiar Services: Professional Footholds in the Entrepreneurial Domain

'Management' as a concept for (self)improvement is very popular in Ahmedabad. The biggest social gathering I attended during a year-long fieldwork in 2001 was a lecture on '*Upanishads*: Golden Secrets for a powerful Personality' by Shri G. Narayana, who was introduced as an 'exponent of Indian wisdom' and who was a consultant in daily life.[2] The event was organized by the Ahmedabad Management Association (AMA). He captured his audience with a mixture of prayers, one-liners and management thoughts about the importance of learning:

God is nothing; out of nothing everything is born. Perfection is inside everybody: you have the power to produce excellence. See the film *Lagaan*: India *can* produce

2 The *Upanishads* are among the oldest Hindu scriptures and focus on learning and self–realization.

quality![3] All we need is a good leader and a good team. Do as Rama did, the perfect worker; do as Krishna did, the perfect manager. God is all vibrations, all consciousness; God is introspection and reaching out to others. AMA is a godly institution if it can reach out and teach us (Narayana, 2001).

Indeed, Ahmedabad has started to think of itself as the 'management capital' of India after the city had witnessed an explosion of private management education in the second half of the 1990s. Nevertheless, the relationship between management studies and the local business community remains rather paradoxical.

The Indian Institute of Management Ahmedabad (IIM-A) is Ahmedabad's oldest business school and one of the most prestigious schools in the whole of Asia. It was set up in collaboration with the Harvard Business School to educate new generations of professional managers for Ahmedabad's textile mills. However, at present its global ties are stronger than its local roots, making it an example of 'de-linking'. Few Gujarati students study there and even fewer students 'stay behind' in Ahmedabad after graduation, not only because they are recruited by reputed Wall Street investment bankers, but also because what they have learnt is simply not being applied by the vast majority of Ahmedabadi companies. For a local lowering of the professional frontier less high–flying professionals are needed.[4]

Whereas the quality of IIM-A's MBA program is globally recognized, the quality of MBA graduates from Ahmedabad's new business schools was doubted, even locally. These schools reportedly hired their own pupils to teach after graduation, despite having no practical business experience.

I attended a workshop entitled, 'Advertising and Sales Promotion: Emerging Challenges', at a local business school that had been opened in 1998, and the morning session was true to form. The first speaker was a medical doctor, who presented on the topic of 'mind management'. His lecture was a strange mixture of 'home-grown' ideas about the functioning of the mind, human psychology and the creation of needs. After his talk, three of the five official guests left. Only I remained with a young sales executive from a local company who had completed a short three-month diploma course in marketing and was hoping to learn something new here.

Following the less successful morning session, the afternoon session demonstrated the relevance of these new business schools. The second speaker was a consultant who discussed sales strategies. He spoke in general terms, using phrases like 'a sales strategy should distinguish a product, create a niche', but his talk was just enough to convince the audience of his experience. This had been his 'share-ware'; it should create demand for more, against a fee. Indeed, after his presentation my fellow official guest approached him to discuss business. For both parties this ramshackle seminar had been worthwhile.

---

3   The film *Lagaan* was nominated for the 2001 Academy Award for best foreign film.

4   See Jonas (2003), who describes a global–professional culture in management consultancy and banking and its relevance for communication between actors from across the globe, groomed as they are in the same culture.

This is a first professional foothold. Consultants like to teach at these new business schools, at technical colleges or, for example, at AMA. Teaching provides them with a regular income but, more importantly, it provides them with a platform and the opportunity to play the role of *guru*. Several consultants referred to the importance of teaching and the role model of the *guru* for positioning themselves, because the older role model of the *'guru'* was more familiar and easier to understand for people than the newer phenomenon of the 'consultant'. What remained unsaid, but was surely part of the equation, was that *gurus* enjoy all the esteem that consultants in Ahmedabad struggled to achieve. Having a platform was also important to consultants as it provided them with a 'neutral' setting where they could demonstrate their expertise without commercial interests playing a direct role, which again added credibility to their presentation.

Aside from serving as a stage for guru role play, local business schools also produce a second professional foothold in the form of cheap middle management, which will *not* leave for Wall Street. This middle management helps to bridge divisions between consultants and entrepreneurs, as was made clear by 'Mrs. Desai', a market research consultant:

> Owners here rule like lord and master. They were doing fine all these years, so why should their authority be challenged with liberalization and globalization? What can I tell them about their customers they do not know by experience? I rarely interact directly with the owners. Most of the time I work with their better-educated staff. They can better appreciate my value. They are like a professional foothold to me (Desai, 2001).

Aside from their (limited) practical skills, the relevance of local MBAs lies in the fact that, as a result of their rudimentary business training, they will start to think in a particular 'thought style', organized around particular concepts and categories, which connects them with other professionals. A three-month diploma course in marketing is just enough to motivate somebody to find his way to a marketing seminar and to appreciate the presentation of a consultant.[5]

Serving as a professional foothold, however, is hardly the professional's dream. When the formal interview was over, Mrs. Desai chose to tell me in confidence about how badly entrepreneurs treated their staff, making local MBAs work without any discretionary powers and forcing them to fetch glasses of water for their boss 'as if they were servants'. This hiring a professional and treating him like a clerk can be seen as a reflection of the struggle to fit a new person into an older system, just like playing the old role of *guru* is a way to make the new role of the consultants more 'fitting' and familiar.[6]

---

5   Douglas (1987) sees a thought style as an imprint of the social and the collectively shared on the individual. A thought style is seen as the precondition for cognition, an individual's pattern of reasoning framed in and working with collectively shared and internalized institutions together making up an individual's thought world.

6   See also Dutta who remarks: 'The upper castes have always invested in technical education, while trading communities have held money-handling skills and marketing in high

During the interview Mrs. Desai also suggested a third professional foothold to access the entrepreneurial domain. Again this foothold is about how 'new', and perhaps less 'tangible', advice on marketing, is being fitted in *via* something more known, better understood, and perhaps also more concrete, such as advertisements:

> The owners distrust my work, thinking it may be all cooked up without any real research being involved. Often they first need to be convinced by their advertisement agency to do some research, instead of just putting it on the market. With advertisement they are a little more familiar than with market research. They also serve as a professional foothold (Desai, 2001).

The most established of these better known, more concrete services in Ahmedabad is chartered accountancy.[7] In a very similar fashion to their big international counterparts, individual chartered accountants (CA) are building on their relatively continuous, trusted relations with clients to become 'service retailers'. 'Mr. Majumbar' went through this process. He began as what he called a 'classical' CA, focusing on taxation and finance. But this was changing in the era of liberalization and globalization:

> Clients' needs now compel me to understand management, human resource development. The concept of providing different services under one roof, like a supermarket, is just developing in India. If my clients require specialists, they go for specialists, but for a simple recruitment procedure they will not go to a human resource development specialist, they will ask me (Majumbar, 2001).

Instead of hiding their business and writing the accounts themselves, clients were becoming 'open-minded' and had begun to ask advice on even 'minor issues':

> Clients have learned that in a margin business they need a CA. Even for small projects clients now come to me for an assessment if the project is viable or not. This is new. Sometimes I find them even becoming over-dependent. They start asking advice on everything (Majumbar, 2001).

Part of building this trust was to go through the same process as young MBAs and literally serve the client. Though Mr. Majumbar also grumbled about clients' unprofessional behaviour, their failure to abide by appointments, behaving like they 'owned' him, he was willing to put up with it as long as progress was made in business. If the client was ready to learn, Mr. Majumbar was ready to learn as well and thus ended up reading up human resource management or finding out about some technical specifications to fill a gap in his own knowledge to please the client.

---

regard ... However, new [business] networks will have to give pride of place to scientists, engineers, technicians and non-business castes' (1997:248).

7   The Ahmedabad branch of the western region (Mumbai) of the Institute of Chartered Accountants India had 2,114 members in April 2000. Membership is obligatory for those who want to practice. Estimations about the number of CAs who had their own practice varied considerably, from 200 to 900.

The local-global interface is not necessarily a single place, a single encounter between (representatives of) different systems. Global standards and institutions and professional best practices can be best understood as travelling through a series of interfaces, from the best-seller management manual or the international consultant located in global cities, to more local research institutions or management associations, to 'ramshackle' seminars and local consultants, and via a number of professional footholds into the entrepreneurial domain. With each step the professional frontier is lowered, but each step also requires translation and adaptation to ensure a local fit. Ahmedabad, though still relatively 'provincial' in nature, is well-connected to the global flow of knowledge, information and skills, seeping in through the obvious channels (Internet, cable television, books), but also in more embodied form: international guest lecturers visiting IIM-A for instance, IIM-A professors lecturing in turn at AMA, local consultants attending those lecturers and lecturing themselves at local business schools. Ideally, a service economy or a knowledge economy harbours a great number of such relatively well-laid out pathways along which exogenous institutions and best practices can travel to reach entrepreneurs. However, despite the mechanisms described here, market dynamics create all but the opposite in Ahmedabad, where consultants occupy the 'service crossroads'.

The Management Consultancy Gap: Monopolized Platforms and Informal Referral in a Fragmented Market

All the brokerage mechanisms described above help to bridge the gap between the professional and the entrepreneurial domain at a *personal* level. In this section, other brokerage mechanisms will be added to this repertoire. At the same time it will show how, at the *market* level what can be labelled the 'reputation game' inhibits the development of such footholds. The way the market for consultancy works produces a consultancy gap. To understand how this market is organized, it is helpful to distinguish five layers of consultants in Ahmedabad.

A first layer consists of large-scale, 'branded', international service providers. These are all Mumbai-based, but have local representatives 'keeping an eye on the Ahmedabad market'. Their main clientele is the Government of Gujarat, because only they are ready to pay the fees for 'branded' advice that has such a standing it is less likely to be questioned in the local political debate. As one local consultant commented: 'It is a battle between the big name with little expertise and the small name with much expertise.' This is a first sign of the 'reputation competition' that shapes the reputation game.

A second layer of consultants is made up of IIM-A faculty members. This is the city's most established layer of consultants. A former director of IIM-A recalled how in the 1980s a consultant was either a fixer, who could get you a bank loan, or a status symbol for established entrepreneurs – something expensive to show off at the Rotary Club. This picture is changing, however, as Professor Mehta's account shows:

I have been doing consultancy for almost 20 years now. In the early days, I had clients from all over India, but not from Ahmedabad. Some ten years ago, the first Ahmedabadi companies came to me. Some five years ago, also small and medium-sized companies started approaching me, because, after liberalization, they wanted to grow. Before, local companies thought that IIM-A was not accessible. They did not expect to find Gujarati speaking faculty members like me here. Also, they were not willing to pay our prices. But people have realized that the traditional way of doing business is simply not good enough anymore. I see a change in attitude. This change is also reflected in the blossoming of AMA. And you can also see it in the arrival of a lot of small consultants, who do a lot of useful systematization in companies (Mehta, 2001).

Despite the arrival of SME at her doorstep, Professor Mehta's clientele mainly consisted of large companies with an average turnover of 100 to 300 *crore*.[8] She, therefore, still operated in the upper segment of the market, but the 'small' consultants she referred to, however, do so much less.

'Small' consultants organized in the Institute of Management Consultants India (IMCI) make up a third layer of consultants in Ahmedabad. They are mostly 'one-man-shows'. The core consists of a first generation of independent consultants in Ahmedabad, who started their practice in the second half of the 1980s and over time have become relatively well-established. The reported minimum turnover of their clientele varies from less than one *lakh* to 25 *crore*, but on average is around one to three *crore*.[9] The minimum value of assignments varies between a few hundred rupees for basic paperwork to studies earning ten *lakh*, with an average of around one *lakh*.

These established consultants face competition from less expensive, less experienced, young or part-time consultants (layer four of the market) and less professional fixers (layer five). Not surprisingly therefore, the aim of IMCI is to professionalize consultancy by means of a diploma course to become an 'ethically operating' Certified Management Consultant, a 'classic' strategy of exclusion.[10] The Ahmedabad chapter of IMCI had around 40 members in 2001, eight of whom were certified. Nevertheless, the IMCI chapter in Ahmedabad was unable to establish itself as a 'safe haven' for professional consultancy: entrepreneurs usually did not know of their existence.[11]

---

8   One *crore* equals ten million. The exchange rate of the Indian Rupee at the time of research was Rupee around 46 to the US dollar.

9   One *lakh* equals 100,000.

10   See, i.e., McDonald (1995) and Murphy (1988). See also Sajor (2003a; 2003b), who describes similar processes of professionalization and exclusion among real estate agents at Metro Cebu in the Philippines.

11   The IMCI secretary estimated a potential membership in Ahmedabad of around 200 members. This, however, does not mean that there are 200 full-time active consultants in Ahmedabad adhering to IMCI standards. Among IMCI's *active* members was an owner of a computer education center who had only *considered* becoming a consultant. Also, among those who were asked to become member were consultants who had admitted to me to be in the business of fake business plans to secure a bank loan. In other words, the number 200 seems

Consultants who belong to layer one and two either enjoy formal employment in reputable consultancy firms or part-time consultancy arrangement under the wings of an institution of equally good reputation. Hence they enjoy the credibility that comes with their position as well as a little 'protection' from market pressures in the form of a regular pay check. On top of this, they normally work in the easier segment of the market: firms with established professional footholds. IMCI consultants enjoy none of these. They depend on their own names and more informal brokerage mechanisms to access a more difficult market segment, in which these footholds are still largely absent. Here the complexity of the market for professional business consultancy in Ahmedabad begins to show and the reputation game really starts to play.

Consultancy is an activity for which no formal quality standards (can?) exist. This means that the market for consultancy services is a market in which it is difficult to distinguish between good and bad quality and, as a consequence, bad quality continuously threatens to drive out good quality.[12] In response, reputation and the posture of professionalism become strategies aimed at distinction. The value of a reputation lies in the fact that a good reputation attracts 'good', 'professional' assignments that will further bolster it, while a lack of reputation will attract 'bad', 'unprofessional' assignments that will further damage it. Reputation helps a consultant to move up or keeps him down.

At the same time, however, consultants are always busy making compromises, simply because the market demands it: (1) 'fixing' remains a necessary consultancy activity along with 'forecasting' in a country plagued by slow administration, complex regulation and corruption; (2) industrial espionage or playing copycat are 'normal' elements of competitive strategy design (e.g., Porter, 1980); and (3) an all-too-professional posture with an emphasis on *replacing* entrepreneurial 'lay knowledge' or 'common sense' with a more 'calculated', 'foresighted' body of knowledge and entrepreneurial autonomy with professional systems may backfire *vis à vis* self-made entrepreneurs (De Swaan, 1988: 244-245). The reputation game is a balancing act between the *posture* of reaching out to the highest standards of *professionalism* and the need to lower standards and *preserve practicality* to be able to work in a non-professional environment.

Maintaining this balance translates in a very conservative approach to the market and genuine shock about the liberal, commercial attitude of international consultants (layer one) who freely advertise their services. 'Mr. Nayar', for instance, was very outspoken in this regard, 'unless one wanted to prostitute oneself', and took pride in

---

to be more an indication of the *total* consultancy population in Ahmedabad, excluding CAs, retired people and part-time consultants and fixers. Numbers for these last two categories are impossible to estimate, because they are open to potentially anyone: retired bankers, employed engineers with a side business, *etcetera*. For customs alone someone estimated between 50 and 80 part–time practitioners versus three full–time consultants.

12 Ackerlof (1970) has made this argument with regard to the market for second-hand cars. One can imagine that with respect to an intangible product such as 'advice', problems of 'quality insecurity' become even bigger.

the fact that he had never printed a brochure, not even a business card. He explained how he had found his first client:

> When I started I was not keen on going anywhere or publishing anything. For three months I was idle, thinking only, approaching nobody. Then a famous Bollywood friend approached me to organize a benefit film premiere to raise funds for cancer research. I took up this opportunity to do social work. While I was looking for sponsors I had the opportunity to meet two or three companies. One company was managed by a young man with a US university degree and full of new ideas. He was impressed by the way in which I managed this event right from planning to execution. He became my first client (Nayar, 2001).

All consultants had these career-defining stories about how they came into contact with their first clients. First clients do not only offer the opportunity to gain experience, but also to gain access to the brokerage mechanism most consultants depend upon: *informal referral*. Once there is a first client who has 'tested the water' and is willing to testify to their credibility, others will follow.

In the absence of formal 'platforms', such as IIM-A functions for its faculty members, independent consultants look for informal stages like business school classrooms. For Mr. Nayar the social event had been one. It had provided him with an opportunity to show his capabilities, but also to show his integrity and moral righteousness by rendering unpaid services (something he stressed), and to meet potential clients without *immediately* having to discuss business (again something he stressed). Non-commercial behaviour in a neutral environment was important. For this purpose Mr. Nayar had also developed another platform: the local Lions Club. Indeed, most consultants were members somewhere.

The next step for a consultant is to develop 'special relationships' and to be added to 'the list' of the platform of choice. Every institution in Ahmedabad has such a list. Enquiries about consultants are channelled to the select number of consultants on that list, but *only informally*. Another step up the ladder is to control an institution or association, in which case you are no longer dependent on others to put you on the list: you *are* the list. It is a strategy to monopolize and 'privatize' the neutral, public and non-commercial image of an organization for commercial means. It is the 'takeover' of a credibility enhancing social institution like the *guru* role model on an organizational level. Such a takeover may explain IMCI's failing public role. An export consultant, 'Mr. Mankad', said he would only become a member if he was 'invited' by the local chairman, 'Mr. Upadhyay', his rival in export consultancy. 'Mr. Nair', also IMCI member, had this to say:

> IMCI is a very dormant organization run by four–five people for their own benefit. I am member, but they never came to me for any kind of help or any kind of cooperation, although I have an excellent track record. Actually the market is small here, so they do not want their clients to be known. Competition is tough, so they want to keep everything secret. These four people tell each other that 'this will be among ourselves only, do not divert this to X, then his name might sell more than ours'. If a client approaches them for something because they represent IMCI, they think 'we grab him, let this client not be

known to another consultant'. Even if they are not very competent in an area, they will not refer the person to a better-qualified consultant (Nair, 2001).

Thus, IMCI is simply less of a public institution that represents the profession as a whole than a privatized vehicle that represents the interests of a few established consultants. It is not exceptional in this regard; other institutions such as the National Centre for Quality Management in Ahmedabad and the Indo-American Chamber of Commerce were run in a similar fashion. Excluded from this list are industrial associations: they in turn are monopolized by established entrepreneurs.

Informal referral and monopolized platforms have another advantage besides being brokerage mechanisms. Circulating in the informal, personal network around certain platforms offers opportunities to check someone's track record and provides the social control that comes with community. In the jungle of Ahmedabad's rapidly expanding business arena, this keeps the world small and surveyable. At the same time, this strategy produces two major disadvantages:

(1) Consultants often seek to monopolize similar platforms, circulate in more or less the same social networks, and, as a result, target the same clientele. A consultant and IMCI member, 'Dr. Dave', confirmed this lack of differentiation among consultants, but gave a strictly financial and reputation-related explanation for it:

> The supply side of consultancy has not yet differentiated; competition among consultants has not yet been internalized. There are reasons for this: with an established entrepreneur, payment will be more, chances of success are higher, and the assignment will likely be more prestigious. With three to five years experience, a consultant already does not touch anything below one *crore* or less. After ten to 15 years, he will not touch anything below fifteen to 30 *crore* (Dave, 2001).[13]

Another more social reason for the lack of differentiation – despite the stiff competition mentioned above by the 'left out' IMCI member – may be that consultants simply do not *meet* other, smaller entrepreneurs, because these tend to shy away from institutions and associations and move in different social circles. *Professionally*, consultants may be brokers of professional tools; *socially* they are not, confining themselves to their own circles and thereby creating few opportunities to meet most entrepreneurs. This is reinforced by the fact that consultants seem far more differentiated professionally than socially. The five professional layers distinguished cannot be matched with five social layers. All consultants seem to have a fairly similar middle class service background. In other words, *Ahmedabad's service sector apparently has yet to follow the recent increase in the number of small-scale industries and the social opening up that it brought to the industrial sector*. The reputation game among consultants only further hinders differentiation. Thus, because of financial, professional and social reasons, *the way the market for consultancy functions creates a consultancy*

---

13 One *crore* is already the official investment limit in plant and machinery for being recognized as an SSI. The large majority of Ahmedabadi companies are SSIs or even smaller, informal units.

gap. This gap results in the most inexperienced, unconnected consultants dealing with the least prestigious but most complex assignments of equally inexperienced, unconnected and not yet professionalized entrepreneurs. This, of course, creates a breeding ground for complaints about fixing and the unreliability of consultants.

(2) The monopolization of IMCI means that the market for consultants is *devoid of a 'neutral' referral service that can broker* between consultant and clients that belong to different circles, and which could mend the consultancy gap.

Finally, given the opaque labyrinth of personal networks and small communities, the absence of formal referral and the level of competition among consultants, they themselves become the final informal brokerage mechanism. This is the least professional strategy to get clients: to cling to whatever passes your corner of the market (unless it is *really* beyond your reach or dignity), brush up on whatever expertise you may lack and, if this is not possible, either refer the client to a business associate and friend, someone of equal status whom you trust to return the favour in due time, or subcontract work to another consultant, for instance a less well-reputed, less-connected part-time or junior consultant, without revealing the client's identity.

Consultants also expand their work terrain by roaming both the broad spectrum of consultancy and the large service hinterland of Ahmedabad stretching hundreds of kilometres into other states in India in search of assignments. But they do so by making use of the same confined social networks described above. It is a *horizontal* expansion of the market that does not necessarily imply a *vertical* 'descent' into other social layers.

Roaming displays the, at times, truly difficult side of being a consultant. Most consultants who belong to the third layer of the consultancy market are high achievers with good grades who gave up secure positions as 'deputy general managers' in respectable companies. Now they operate from home or from small, sometimes sparsely furnished offices, fiddle with the air conditioner during an interview to save energy, and travel halfway across the country *by train* for assignments. 'Dr. Dave', for instance, had a background teaching in the reputed Xavier Institute of Management in Mumbai and had done consultancy for the UN, but worked from a small home office in an undistinguished apartment complex and regularly travelled by train across India because that was where he had developed a market. At the same time, another consultant, 'Mr. Purohit', regularly travelled for the same reason in the opposite direction, to Ahmedabad. Like an MBA, being a consultant was much less glamorous than the professional image might suggest.

Conclusions: A Service Market in Development

The role of consultants in the local economic development of Ahmedabad is ambiguous. Their ability to *lower* the professional frontier and introduce exogenous institutions and best practices for institutional compatibility is offset by the reputation game, the need to *move up* to more professional clients and more sophisticated assignments. Their ability to *broker together* the pathways *via* which exogenous institutions and best practices can travel to the Ahmedabad business community is offset by the need to *monopolize* corners of the market, thereby reducing it to a fragmented labyrinth of informal referral. The outcome is equally ambiguous: a service configuration that looks barren at the surface, with low-quality business schools, with better-quality institutions only informally referring consultants to clients, and with an obscure consultancy market populated with, at times, equally obscure consultants; but a service configuration that is also more sophisticated than might be expected, with informal networks reaching far and with many interesting developments taking place, albeit on a small, experimental scale.

One can find successful CAs turning from being passive providers of statutory services into proactive matchmakers, even translators, between their local clients and Mumbai-based companies in order to bring together 'highly professionally formulated demands and local capabilities'. One can also find intricate networks consisting of a variety of consultants *and* entrepreneurs, with the latter in a double role, as receivers of consultancy services for their own projects and as providers of technical support to others, all organized with minimum overhead in a conscious effort to right size services rather creating ill-fitting under or over–professionalization. One can also find, however, small entrepreneurs, who feel the pinch of competition and search for information, but who have no clue as to where to go. The local association is monopolized by larger-scale competitors, local research institutions are not interested in small entrepreneurs, local management associations only organizing lectures in English, and all organizations refuse to refer 'unconnected' entrepreneurs to their informal batch of consultants.

There are reasons to believe that the gap between the professional and the entrepreneurial domain might have been relatively bigger in Ahmedabad at the time of research, but will narrow down over time:

1. Being a centre of trade and industry for generations, Ahmedabad has a distinct entrepreneurial culture built around an almost 'Protestant' ethos of frugality and family. There is a long-standing preference for using a nephew to sort things our rather than bringing in a 'stranger'. Other cities, with less developed indigenous business institutions might be more receptive to exogenous learning. However, also in Ahmedabad consultants are gaining ground as competition intensifies and margins narrow down.

2. Ahmedabad's entrepreneurial community consists of a relatively large number of small, first-generation entrepreneurs, often less-educated, self-made men who manage their business single-handedly and who have little regard for diplomas and specialists. Second-generation entrepreneurs, however, already tend to be better-educated than their fathers and hence are more receptive to 'foreign' best practices and consultancy services.

3. India's economy is in transition from a regulated to a more liberal economy, which has implications for the demand and supply of services. Demand for competition-related services in the field of efficiency, quality and marketing

is relatively new compared to demand fore regulation-related services such as tax consulting. Also, the first-generation suppliers of these services often have a background in India's large-scale industries and may be less familiar with the needs of Ahmedabad's new SSIs. However, one already can find second-generation consultants already more habituated to working with them, more sensitive to the problems they face, their allergy for paperwork and the need to make things practical and applicable.

4. A dynamic city 'out of control' is unlikely to be a good place to find well-developed, fine-turned knowledge networks. Clustering requires a common vision supported by social capital and both take time to materialize.

There are, however, also signs which indicate that this gap has structural features which will not easily wither over time and will also be noticeable elsewhere:

1. Professionalism is about trying to *replace* 'lay knowledge' or 'common sense' with a more 'calculated', 'foresighted' body of knowledge and 'gut feeling' with professional systems. Although management is a way to objectify the art of entrepreneurship, it will not succeed in fully replacing it. Finding the right balance is a continuous struggle, which sophisticated delivery mechanisms for services can help to minimize by offering the right solution at the right time, but cannot prevent.

2. Knowledge, information and skills are risky commodities to buy. Knowledge is 'sticky' and might be difficult to apply outside its original context. Information is quickly outdated and often too generic in nature to be of use. Skills, or practices, are complex in design. Also, all are intangible goods, which are relatively easy to copy (which may be the far more attractive thing to do rather than investing in a new and risky process of innovation, unless intellectual property rights are enforced). Also the consultancy business operates without strong quality standards, which creates opportunities for bad quality to undercut good quality. All this shapes the market and turns it into a reputation game. Such reputation games are also played outside Ahmedabad. Glückler and Armbrüster (2003) describe a similar reputation and network-based market ordering in Europe. Also, the fact that at the international level a small number of 'trusted' companies deliver a wide range of services should be understood as a way to work around risk by investing in reputation. In a city like Mumbai one can see these international companies incorporating independent consultants and, in this way, slowly taking over the upper and middle segment of the market. In the Ahmedabad market, consisting mainly of SSI, such a company-based 'ordering' of the market might never be possible, because here companies with large overhead will never be able to sustain themselves. The alternative is more informally organized consultancy networks, if possible supported by formal referral and neutral platforms to meet clients (as opposed to informal referral and monopolized platforms). Indeed, few such platforms can be found in Ahmedabad. They, for instance,

take the shape of associations and chambers not controlled by entrepreneurs, but by professional staff with the autonomy to set the agenda. They function as 'controlled' market places where consultants and entrepreneurs can meet (see Bekkers, 2005, Bekkers, in preparation).

To conclude, Ahmedabad's service market in development is unique, but at the same time it shares two key characteristics with service markets in many other emerging metropolitan cities in the Asia-Pacific region. Part of a global hierarchy of cities and relatively well-connected to the global flow of services, and pressurized by the need for institutional compatibility and convergence, all these cities face the challenge to lower the professional frontier and develop delivery mechanisms for services that minimize the risk inherent in the commodities involved while ensuring a local fit. In Mumbai global service providers may be instrumental in this, in Penang it may be the local public development corporation and in Ahmedabad it may be informal networks around consultants (Visscher, forthcoming). Delivery mechanisms may differ, the challenge is there for all.

References

Ackerlof, George (1970), 'The market for 'lemons': quality uncertainty and the market mechanism', *The Quarterly Journal of Economics*, **84** (3), August, 488–500.

Bekkers, Harald (2005), 'Between fixing and forecasting: provincial Ahmedabad brokered into a bridgehead for globalization from below', unpublished Ph.D. thesis, University of Amsterdam.

Bekkers, Harald (in preparation), 'Brokering globalization from below in provincial Ahmedabad: the need for providers regulating intangible services and competing interests', in Visscher, Sikko and Muijzenberg, Otto van den (eds), *Globalizing Cities, Local Practices: Analyses of Globalization from Below in Modern Asia, 1960–2000*.

Bessant, J. (1991), *Managing Advanced Manufacturing Technology: The Challenge of the Fifth Wave*, Oxford: Blackwell Publishers.

Best, Michael H. (1990), *The New Competition: Institutions of Industrial Restructuring*, Polity Press: Cambridge.

Dave, P.V. (2001), Personal communication, 16 July, Ahmedabad.

Desai, P.S. (2001), Personal communication, 18 July, Ahmedabad.

De Swaan, Abram (1988), *In Care of the State: Healthcare, Education and Welfare in Europe and the USA in the Modern Era*, Cambridge and New York: Polity Press and Oxford University Press.

Douglas, Mary (1987), *How Institutions Think*, London: Routledge and Kegan Paul.

Dutta, Sudipt (1997), *Family Business in India*, New Delhi, Thousand Oaks and London: Response Books/Sage.

Fligstein, Neil (2001), *The Architecture of Markets: An Economic Sociology of Twenty-First-Century Capitalist Societies*, Princeton and Oxford: Princeton University Press.

Glückler, Johannes and Thomas Armbrüster (2003), 'Bridging uncertainty in management consulting: the mechanisms of trust and networked reputation', *Organization Studies*, **24** (2), 269–297.

Government of Gujarat (2001), *Socio-Economic Review Gujarat State 2000–2001*, Gandhinagar: Directorate of Economics and Statistics.

Jonas, Andrew (2003), *Management Consultancy and Banking in an Era of Globalization*, Basingstoke and New York: Palgrave Macmillan.

Lakha, Salim (1988), *Capitalism and Class in Colonial India: The Case of Ahmedabad*, Asian Studies Association of Australia, South Asia Publication Series, No. 3, New Delhi: Sterling Publishers.

Majumbar, R.V. (2001), Personal communication, 25 June, Ahmedabad.

McDonald, Keith M. (1995), *The Sociology of the Professions*, London, Thousand Oaks and New Delhi: Sage.

Mehta, I (2001), Personal communication, 5 November, Ahmedabad.

Murphy, Raymond (1988), *Social Closure: The Theory of Monopolization and Exclusion*, Oxford: Clarendon Press.

Nair M.M. (2001), Personal communication, 21 June, Ahmedabad.

Narayana, Shri G. (2001), '*Upanishads*: golden secrets for a powerful personality', lecture at the Ahmedabad Management Association, 29 August, Ahmedabad.

Nayar, A. (2001), Personal communication, 3 March, Ahmedabad.

North, Douglass C. (1985), 'Transaction cost in history', *Journal of European Economic History*, **14** (3), 557–576.

North, Douglass C. (1990), *Institutions, Institutional Change and Economic Performance*, Cambridge: Cambridge University Press.

Oonk, Gijsbert (1998), *Ondernemers in Ontwikkeling: Fabrieken en Fabrikanten in de Indiase Katoenindustrie*, Hilversum: Verloren.

Porter, Michael E. (1980), *Competitive Strategy: Techniques for Analyzing Industries and Competitors*, New York, Toronto, Oxford, Singapore and Sydney: The Free Press.

Porter, Michael E. (1990), *The Competitive Advantage of Nations*, New York, London, Toronto, Sydney and Singapore: The Free Press.

Sajor, Edsel (2003a), 'Land investors during the property boom of the 1990s and the elite of Metro Cebu', in Dahles, Heidi and Muijzenberg, Otto van den (eds), *Brokers of Capital and Knowledge: Changing Power Relations*, London and New York: RoutledgeCurzon.

Sajor, Edsel (2003b), 'Globalization and the urban property boom in Metro Cebu, Philippines', *Development and Change*, **34** (4), September, 713–741.

Spodek, Howard (1965), 'The manchesterization of Ahmedabad', *Economic Weekly,* **17**, March 13, 483–490.

Spodek, Howard (1969), 'Traditional Culture and Entrepreneurship: A Case Study of Ahmedabad', *Economic and Political Weekly*, **4** (8), February 22, Review of Management, M27-M31.

Spodek, Howard (1989), 'From Gandhi to violence: Ahmedabad's 1985 riots in historical perspective', *Modern Asian Studies*, **23** (4), October, 765–795.

Spodek, Howard (2001), 'Crises and response: Ahmedabad 2000', *Economic and Political Weekly*, **36** (19), May 12, 1627–1638.

Visscher, Sikko (forthcoming), 'Constructing Penang: globalization, modernity and lifestyle in the real estate development sector in Penang during the NEP and Vision 2020 period', in Visscher, Sikko and Muijzenberg, Otto van den (eds), *Globalizing Cities, Local Practices: Analyses of Globalization from Below in Modern Asia, 1960–2000*, London and New York: RoutledgeCurzon.

Index

(References to maps and figures are in bold)

For Product Safety Concerns and Information please contact our
EU representative GPSR@taylorandfrancis.com Taylor & Francis
Verlag GmbH, Kaufingerstraße 24, 80331 München, Germany